I0484387

EPA/600/R-04/179
January 2005

Monitored Natural Attenuation of MTBE as a Risk Management Option at Leaking Underground Storage Tank Sites

John T. Wilson, Philip M. Kaiser and Cherri Adair -
U.S. Environmental Protection Agency
Office of Research and Development
National Risk Management Research Laboratory -
Ada, Oklahoma 74820 -

Project Officer -
John T. Wilson -
Ground Water and Ecosystems Restoration Division -
National Risk Management Research Laboratory -
Ada, Oklahoma 74820 -

National Risk Management Research Laboratory -
Office of Research and Development -
U.S. Environmental Protection Agency -
Cincinnati, OH 45268 -

Notice

The U.S. Environmental Protection Agency through its Office of Research and Development funded the research described here. It has been subjected to the Agency's peer and administrative review and has been approved for publication as an EPA document. Mention of trade names and commercial products does not constitute endorsement or recommendation for use.

All research projects making conclusions and recommendations based on environmentally related measurements and funded by the U.S. Environmental Protection Agency are required to participate in the Agency Quality Assurance Program. This project was conducted under two Quality Assurance Project Plans. Work performed by U.S. EPA employees or by the U.S. EPA on-site analytical contractor followed procedures specified in these plans without exception. Information on the plans and documentation of the quality assurance activities and results is available from John Wilson or Cherri Adair.

Foreword

The U.S. Environmental Protection Agency is charged by Congress with protecting the Nation's land, air, and water resources. Under a mandate of national environmental laws, the Agency strives to formulate and implement actions leading to a compatible balance between human activities and the ability of natural systems to support and nurture life. To meet this mandate, EPA's research program is providing data and technical support for solving environmental problems today and building a science knowledge base necessary to manage our ecological resources wisely, understand how pollutants affect our health, and prevent or reduce environmental risks in the future.

The National Risk Management Research Laboratory (NRMRL) is the Agency's center for investigation of technological and management approaches for preventing and reducing risks from pollution that threatens human health and the environment. The focus of the Laboratory's research program is on methods and their cost-effectiveness for prevention and control of pollution to air, land, water, and subsurface resources; protection of water quality in public water systems; remediation of contaminated sites, sediments and ground water; prevention and control of indoor air pollution; and restoration of ecosystems. NRMRL collaborates with both public and private sector partners to foster technologies that reduce the cost of compliance and to anticipate emerging problems. NRMRL's research provides solutions to environmental problems by: developing and promoting technologies that protect and improve the environment; advancing scientific and engineering information to support regulatory and policy decisions; and providing the technical support and information transfer to ensure implementation of environmental regulations and strategies at the national, state, and community levels.

In the United States of America, the responsibility for managing spills of gasoline from underground storage tanks falls to the individual states. Where it has been appropriate, many states have selected monitored natural attenuation as a remedy for organic contaminants in ground water. Many states also use a formal process of risk management to select the most appropriate remedy at gasoline spill sites. Both monitored natural attenuation (MNA) and risk management require an understanding of the environmental processes that control the behavior of a contaminant in ground water. This report is intended for technical staff in the state agencies with responsibility for administering the underground storage tank program as mandated by RCRA. The information is intended to allow the state regulators to determine whether they have adequate information to evaluate MNA of fuel oxygenates at a site, and to allow the regulators to separate sites where MNA of fuel oxygenates may be an appropriate risk management alternative from sites where MNA is not appropriate.

Stephen G. Schmelling, Director
Ground Water and Ecosystems Restoration Division
National Risk Management Research Laboratory

Abstract

This report reviews the current state of knowledge on the transport and fate of MTBE in ground water, with emphasis on the natural processes that can be used to manage the risk associated with MTBE in ground water or that contribute to natural attenuation of MTBE as a remedy. It provides recommendations on the site characterization data that are necessary to manage risk or to evaluate monitored natural attenuation (MNA) of MTBE, and it illustrates procedures that can be used to work up data to evaluate risk or assess MNA at a specific site.

Contents

Figures

Tables

Acknowledgments

Douglas Mackay at the University of California at Davis, Michael Hyman at North Carolina State University, Theresa Evanson and Aristeo Pelayo with the Wisconsin Department of Natural Resources, and Harold White with U.S. EPA Office of Underground Storage Tanks provided formal peer reviews. Tomasz Kuder at the University of Oklahoma, Ravi Kolhatkar with Atlantic Richfield Company, Patricia Ellis with the State of Delaware Department of Natural Resources and Environmental Control, and Matthew Small with U.S. EPA Region 9 provided technical reviews. Seth Daugherty with the Local Oversight Program within the Environmental Health Division of the Health Care Agency of Orange County, California, provided technical reviews and valuable suggestions for the technical approach taken in the document and the format and organization of the document.

Section 1

Relationship Between Risk Management and MNA -

Monitored natural attenuation (or MNA) is defined by U.S. EPA in the OSWER Directive (U.S. EPA, 1999) as one *alternative means of achieving remediation objectives that may be appropriate for specific, well-documented site circumstances where its use meets the applicable statutory and regulatory requirements.* The remedial objective may be chemical-specific cleanup levels. The remedial objective may also include *preventing exposure to contaminants, preventing further migration of contaminants from source areas, preventing further migration of the groundwater contaminant plume, reducing contamination in soil or groundwater to specified cleanup levels appropriate for current or potential future uses, or other objectives.*

Natural attenuation processes, such as biodegradation and dispersion along a flow path, can bring the concentration of contaminants to a chemical-specific cleanup level. This is particularly true when the source of contamination has been controlled. The same natural attenuation processes can also prevent exposure to contaminants, prevent further migration of contaminants from source areas, or prevent further migration of the ground water contaminant plume.

This section discusses the relationship between risk management and MNA, describes the common remedial objectives for monitored natural attenuation, identifies the behavior of ground water plumes that is crucial for success in monitored natural attenuation, and makes suggestions to improve the current state of practice for monitored natural attenuation of methyl tertiary butyl ether (MTBE).

1.1 Monitored Natural Attenuation and Risk Management

In the United States of America, the responsibility for managing spills of gasoline from underground storage tanks falls to the individual states. Where it has been appropriate, many states have selected monitored natural attenuation as a remedy for organic contaminants in ground water (U.S. EPA 1999; New England Interstate Water Pollution Control Commission; 2000, 2003).

Many states use a formal process of risk management to select the most appropriate remedy at gasoline spill sites. The potential receptors of contamination are identified, and the behavior of the plume is characterized to determine the potential for contamination to migrate along a flow path and impact the receptors. Many states estimate that risk with mathematical formulas or mathematical models that describe the rate of transport of contaminants in ground water and the rate of attenuation of the contaminant along the flow path through dilution and dispersion, sorption, and biodegradation.

If the remediation objective for MNA is to prevent exposure or to prevent further migration of MTBE, the critical issue is the distance MTBE can move in ground water. The size and long-term behavior of a plume of MTBE are controlled by the rate of dissolution of MTBE from the residual gasoline in the source area and the rate of attenuation along the flow path in the aquifer through biodegradation, dilution, and dispersion (Small and Weaver, 1999). This interaction is described and illustrated in detail in *Section 2 Typical Behavior of MTBE Plumes.*

If the remedial objective is a specific cleanup level, the critical issue is the time required to reach the cleanup goal. In most gasoline spills, the long-term source of MTBE in ground water is MTBE in residual gasoline trapped in the aquifer. The time required to reach the clean up goal is controlled by the rate at which MTBE dissolves from the residual gasoline into ground water as the ground water flows past the residual gasoline. If the MTBE dissolves rapidly, the residual gasoline will be depleted of MTBE. As ground water from up gradient comes into contact with the depleted gasoline, the concentration of MTBE that dissolves into the ground water will be less. The concentrations of MTBE in water will drop rapidly. If the MTBE dissolves slowly, residual gasoline will be depleted slowly, and MTBE will persist in the source area for long periods of time.

The OSWER Directive (U.S. EPA, 1999) requires that MNA will *meet site remediation objectives within a time frame that is reasonable compared to that offered by other methods.* The progress toward achieving the remedial objective is monitored until the remedial objective is obtained. When the remedial objective is to prevent exposure and prevent further migration of contaminants, many states will monitor the concentrations of contaminants until a statistical analysis of the data reveals that the concentrations are declining over time at some predetermined level of confidence. When the remedial objective is a specific cleanup level, many states will monitor a site until the cleanup level is met.

1.2   Suggestions to Improve Plume Management and Risk Evaluation

There is room for improvement in the current practice for risk evaluation of MTBE plumes in ground water. Most MTBE plumes are anaerobic. Until recently, it was generally believed that MTBE would not degrade in anaerobic ground water, and most risk assessments for MTBE have ignored the possibility that the MTBE might biodegrade. Recent work has documented anaerobic MTBE biodegradation in laboratory microcosm studies. *Section 3 Recommendations for Monitoring* discusses the monitoring needed to recognize and properly describe anaerobic MTBE biodegradation at field scale. *Section 4 Biological Degradation of MTBE and Other Fuel Oxygenates* reviews the current state of knowledge concerning the microbiology of MTBE biodegradation.

In the past, the primary evidence for biodegradation at field scale was the accumulation of degradation products. The primary degradation product of MTBE is tertiary butyl alcohol (TBA). Because TBA has intentionally been added to gasoline as a fuel oxygenate, and because it occurs as a trace component of commercial MTBE in gasoline, TBA accumulation by itself is not convincing evidence of MTBE biodegradation. This makes it particularly difficult to use conventional monitoring data to document biodegradation of MTBE at field scale, or to extract rate constants for attenuation that can be used in predictions of the future behavior of plumes.

Recent work has shown the stable carbon isotopes in MTBE are fractionated when MTBE is biologically degraded (Hunkeler et al., 2001; Gray et al., 2002; Kolhatkar et al., 2002; Kuder et al., 2005). As biodegradation proceeds, the MTBE that has not been degraded has a progressively greater proportion of the heavy carbon isotope ^{13}C, compared to the more common isotope ^{12}C. Advances in compound-specific stable isotope analyses make it possible to accurately measure the shift in the ratio of the isotopes in MTBE in water at low concentrations. The fractionation of the MTBE that has not degraded becomes the equivalent to a "metabolic product" that is used to document biodegradation. This makes it possible for the first time to unequivocally identify and measure anaerobic biodegradation of MTBE at field scale. *Section 5 Monitoring MTBE Biodegradation with Stable Isotope Ratios* explains the units used to measure carbon isotope fractionation and discusses the simple formulas used to estimate the extent of biodegradation from the extent of fractionation. *Section 6 Application of Stable Isotope Ratios to Interpret Plume Behavior* illustrates the use of stable carbon isotope analyses to recognize anaerobic biodegradation of MTBE at field scale, to extract a rate constant of biodegradation of MTBE at field scale, and to evaluate the contribution of MTBE biodegradation to the concentrations of TBA measured at a gasoline spill site.

1.3   Suggestions to Improve MNA to Meet a Cleanup Goal

There is also room for improvement in the current practice to evaluate the rate of natural attenuation over time. Often the monitoring data are presented to state regulators as a simple chart or table without any statistical evaluation of the data. If the data are examined, the evaluation is often cursory and incomplete. The rate of attenuation over time is conventionally estimated as the slope of a linear regression of the monitoring data on the date of sampling. The report may provide the regulator with a chart showing the regression line. It may also provide the correlation coefficient (r^2) of the regression as an indication of the variability of the data. The value of r^2 in itself is not a test for statistical significance. The monitoring data should be evaluated to determine if the concentrations are actually declining. More specifically, the data should be evaluated to determine if the slope of the regression line is statistically significant from zero at some predetermined level of confidence.

Section 7 Statistical Evaluation of Rates of Attenuation of Sources provides detailed step-by-step instructions to extract a rate of attenuation from field data and to evaluate the data to determine whether the rate is statistically significant from zero. The approach that is illustrated relies on conventional parametric statistics. It is the approach that is most likely to be familiar and accessible to a ground water scientist or engineer that does not have extensive experience with environmental statistics.

Section 8 Typical Rates of Attenuation in Source Areas presents data on typical rates of attenuation in the source area of selected MTBE plumes. The typical rates provide a benchmark for the relative rate of attenuation at a particular site of interest. They also provide a realistic view of the prospects for rapid natural attenuation of MTBE in the source area of plumes in general. This section also provides information on the number of samples that are typically necessary to determine a rate of attenuation that is statistically significant.

Section 2 -

Typical Behavior of MTBE Plumes -

This section describes the maximum concentrations of MTBE in the source areas of plumes from gasoline spills. The concentration of MTBE in the source area has a strong influence on the length of a plume at steady state and on the prospects that the concentrations will decline to meet a goal for cleanup. This section also describes the typical behavior of the plumes in terms of the rate of natural decline in concentrations of MTBE over time in the source area and the rate of natural decline in concentration of MTBE in ground water along the flow path. The decline in concentration along the flow path is the critical behavior of a plume that determines its ability to impact a receptor. The decline in concentration over time is the critical behavior of a plume that determines its ability to reach a cleanup goal.

2.1 Concentration of MTBE Expected in Gasoline Spills

The concentrations of MTBE in ground water at gasoline spill sites are much lower than would be expected from typical concentrations of MTBE in gasoline. Figure 2.1 compares the distribution of the maximum concentration of MTBE in monitoring wells in southern California, Texas, and Kansas. Oxygenated gasoline is not required in Kansas; however, it is required in California and in the Dallas/Fort Worth and Houston markets in Texas. The data from Texas are the maximum concentrations of MTBE at 609 gasoline spill sites that had at least one analysis for MTBE in monitoring wells at the site (Mace and Choi, 1998). The data from California are the maximum MTBE concentrations at gasoline spill sites in Orange County, California, in 2002 (data courtesy Seth Daugherty, Orange County Local Oversight Program, compare Odencrantz, 1998), and in Los Angeles County, California, in 2002 (Shih et al., 2004). The data from Kansas are the maximum concentration reported at sites in Kansas UST trust fund 2003 (Hattan et al., 2003). The data presented in Figure 2.1 are the maximum concentrations of MTBE in any well at the site within a particular year of monitoring. They are not the maximum concentrations that have ever been recorded at the sites.

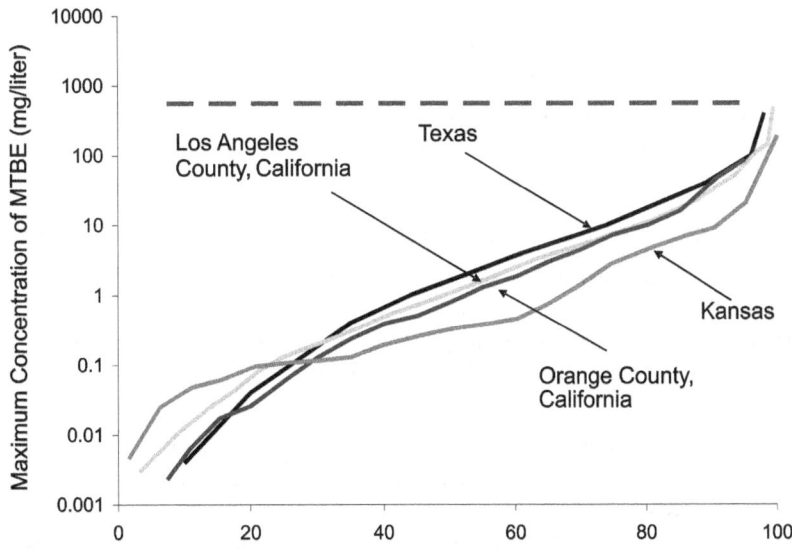

Figure 2.1 *Comparison of the distribution of the maximum concentration of MTBE at gasoline spill sites in regions of the United States that used MTBE to meet the federal oxygenate standard (Texas and Southern California) to a region that did not (Kansas). The dashed line represents the concentration of MTBE that would be expected in ground water in contact with residual gasoline at 1,000 mg/kg when the gasoline contained 11% MTBE. The concentrations of MTBE are similar in regions that used MTBE as the fuel oxygenate. In a region that does not require a fuel oxygenate, the concentrations are lower. The concentrations are less than would be expected based on dissolution of MTBE from reformulated gasoline.*

The frequency distribution of MTBE in ground water in the three regions of the United States that require an oxygenate in gasoline was very similar. Approximately 25% of sites had concentrations less than 0.1 mg/L, the median concentration was near 1 mg/L, approximately 25% of sites had concentrations above 10 mg/L, and approximately 5% of sites had concentrations above 100 mg/L. The frequency distribution of concentrations of MTBE in ground water in Kansas was similar to the distribution in Texas and southern California, but the concentrations of MTBE in ground water in Kansas are on the order of three-fold less than concentrations in these areas where reformulated gasoline is required (Figure 2.1).

The equilibrium partitioning of MTBE between gasoline and water was used to provide a basis for comparison between the actual concentration of MTBE in ground water and the concentration that would be expected from a spill of reformulated gasoline.

$$C_w = \frac{C_{o,NAPL}}{K_{NAPL} + \dfrac{\theta_w}{\theta_{NAPL}}}$$

Equation 2.1

Equation 2.1 calculates the concentration of MTBE in the ground water C_w, from the concentration of MTBE in the gasoline that was spilled $C_{o,NAPL}$, the porosity filled with gasoline θ_{NAPL}, the water-filled porosity θ_w, and the distribution coefficient between gasoline and water K_{NAPL}. The specific gravity of gasoline is near 0.78, which is equivalent to 780,000 mg/L. If the gasoline contains 11% MTBE by volume, then the concentration of MTBE in the gasoline $C_{o,NAPL}$ is 85,800 mg/L. Assume the concentration of residual gasoline in the aquifer sediment is 1,000 mg/kg, and the bulk density of the sediment is 1.7 kg/L. The concentration of gasoline in a volume of aquifer material would be 1,700 mg/L. If the density of gasoline is 780,000 mg/L, the porosity filled with gasoline θ_{NAPL} is 0.00218 L/L. At a bulk density of 1.7 kg/L, the total porosity is 30% of the volume of the aquifer material. Because the porosity filled with gasoline is so small, the water-filled porosity θ_w will be near 0.30. The gasoline to water partition coefficient of MTBE is assumed to be 16 (Cline et al., 1991; Rixey and Joshi, 2000).

$$C_W = \frac{85,800\,mg/L}{16 + \dfrac{0.30}{0.00218}}$$

Equation 2.1 solved

Under these assumptions, the predicted concentration of MTBE in ground water in contact with the residual gasoline is 560 mg/L. This prediction is the horizontal dashed line in Figure 2.1.

The assumed concentration of residual gasoline in the aquifer was 1,000 mg/kg TPH. This is a relatively low value for residual gasoline at spill sites. Most gasoline spills would have higher concentrations of gasoline at residual saturation which would produce higher concentrations of MTBE in water. The measured concentrations of MTBE in the most contaminated monitoring wells in Texas and southern California are from two to three orders of magnitude less than the concentrations that would be expected from the content of MTBE in oxygenated reformulated gasoline.

There are a number of explanations why the measured concentrations of MTBE are lower than the expected concentrations. Not every gasoline spill in these data sets contained 11% MTBE. Many sites have been subjected to active remediation. In some sites, the rate of dissolution of MTBE from the residual gasoline to ground water is limited by mass transfer phenomena. As a consequence, the MTBE in the residual gasoline is not in equilibrium with the MTBE in water.

Because the measured concentrations of MTBE at gasoline spill sites are so much lower than the expected concentrations, the prospects of reaching clean up goals are more attainable. The length of a plume at equilibrium is related to the concentration at the source. Because the concentrations are lower, the possibility of impacting a receptor is less. The lower range of the U.S. EPA health advisory is 20 μg/L. If half the MTBE plumes in those areas of the United States that used MTBE as a fuel oxygenate have a maximum concentration of 1,000 μg/L or less (see Figure 2.1), then half the plumes in their source areas are within a factor of fifty or less of the lower range of the U.S. EPA health advisory. If 25% have a maximum concentration of 100 μg/L or less (Figure 2.1), then 25% are within a factor of five or less of the lower range.

Kansas is probably representative of the regions of the United States where MTBE is not intentionally added to gasoline to meet the federal oxygenate standard. However, it should not be surprising to find MTBE in gasoline spills in

Kansas. Gasoline may contain MTBE for a variety of reasons. The gasoline may have been refined for another market where an oxygenate is required, or MTBE may have been added to meet octane requirements for the fuel, or MTBE may have entered the gasoline through incidental blending with other products or feed stocks that contained MTBE during refining and distribution.

To determine the amount of MTBE in the fuel supply in Kansas, the Kansas DHE analyzed 1,380 fuel samples for the content of MTBE. Only 5.3% of samples had MTBE concentrations in a range from 11% to 15.4% as would be expected for oxygenated reformulated gasoline. Samples of gasoline that contained MTBE for purposes of meeting octane requirements were more common; 21.3% of samples had concentrations of MTBE between 11% and 6%, and 37.6% of the gasoline samples had concentrations of MTBE between 6% and 1.5%. Gasoline that had MTBE from incidental activities were also common; 35.8% of the samples of gasoline had less than 1.5% MTBE (Hattan et al., 2003). The average concentration of MTBE over all the samples of gasoline from Kansas was near 4.5%, between a half and a third of the concentration expected in oxygenated reformulated gasoline. The concentration of MTBE in ground water at gasoline spill sites in Kansas was between one-half and one-third of the concentrations in Texas and southern California where oxygenated reformulated gasoline was required (Figure 2.1).

2.2 Definitions and Expressions Used for Rate Constants

At field scale, the concentration of MTBE in a well can change through the combined influence of dilution and dispersion, biodegradation, sorption, and mixing of the contaminant plume with cleaner water in a monitoring well. It is usually impossible to separate the individual contributions of each process. It is important to acknowledge this uncertainty. In this document, rate constants calculated from changes in concentrations in wells will be identified as rates of attenuation. In laboratory studies conducted with batch microcosms, there is no opportunity for dilution and dispersion, and the effects of sorption are evaluated in sterilized controls. In this document, rate constants that are calculated from controlled laboratory studies will be termed rates of biodegradation.

Dispersion and dilution of MTBE with the flow of water is independent of the concentration of MTBE. As a consequence, their effect on concentrations of MTBE will be proportional to the initial concentration of MTBE. A constant quantity of MTBE does not partition between water and residual gasoline. Rather, a constant proportion of the MTBE will partition, regardless of the absolute concentration of MTBE.

At the concentrations of MTBE most commonly seen in ground water, the rate of biodegradation is not a fixed number, but is proportional to the concentration of MTBE present in the ground water. However, an exception to this general rule will be discussed in *Section 4 Biological Degradation.*

Chemists and engineers describe a process where the rate of the process is directly proportional to the amount of material subject to the process as being a first order process. A first order process for biodegradation is quantitatively defined by Equation 2.2, where F is the fraction of the original material remaining at some time t, and k is the first order rate of removal through biodegradation or attenuation. F is conventionally calculated as the concentration remaining C divided by the original concentration Co.

$$F = C/Co = e^{-kt}$$

Equation 2.2

Equation 2.2 describes changes in concentration over time in a particular monitoring well. Equation 2.3 uses the same relationship to describe changes in concentration with distance along a flow path in the aquifer, where d is the distance along the flow path between the up gradient well producing water with the contaminant at concentration Co and the down gradient well producing water with the contaminant at concentration C.

Equation 2.3

$$F = C/Co = e^{-kd}$$

The most familiar example of a first order process is the decay of radioactive elements. A familiar unit for the rate of radioactive decay is the half life of the element, the time required for half the material originally present to decay. From Equation 2.2, this would be the time required, at a particular first order rate of removal, for C/Co to equal ½. A first order rate constant is not intuitively obvious to most people. Most people find a half life easier to understand. However, a first order rate constant is a direct expression of the rate and is the best way to compare rates with each other. As the rate goes up or down, the value of the constant goes up or down proportionately. First order rate constants are also more convenient to use in equations. As a consequence, the first order rate constant will be used throughout this manuscript. If a reader prefers to think in terms of half lives, Figure 2.2 provides a convenient means to translate a first order rate constant into a half life. A half life is easily calculated from a first order rate constant using the relationship in Equation 2.4, where t is the time required to reach half the initial concentration, and k is the first order rate constant for attenuation. The unit of the half life is the reciprocal of the unit for the first order rate constant.

$$t_{1/2} = 0.693/k$$ Equation 2.4

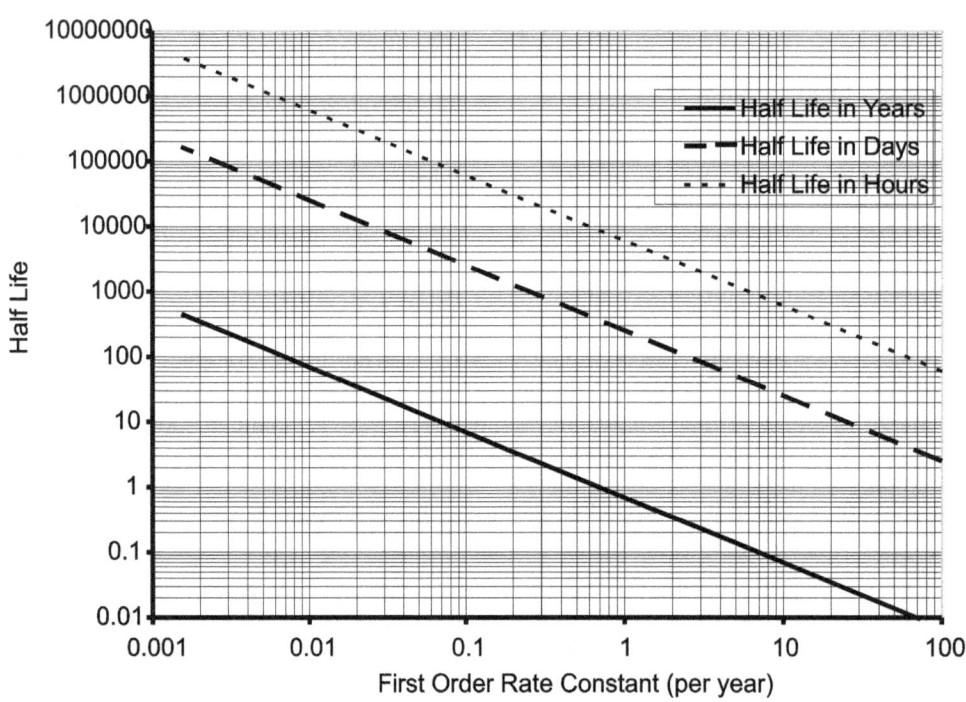

Figure 2.2 *Relationship between a first order rate constant and a half life.*

2.3 Role of Attenuation in the Lifecycle of MTBE Plumes

The size of a plume reflects a balance between the rate of release of the contaminant from the source area into the aquifer, the rate of transport of the contaminant away from the source area, and the rates of degradation and dispersion in ground water which remove mass or reduce plume concentrations. Depending on the relationship between the rate of dissolution of the contaminant from the fuel spill and the rate of attenuation of the contaminant in ground water, plumes may grow, or plumes may shrink and eventually disappear. When source dissolution and advection dominate over dispersion and biodegradation, plumes expand. When biodegradation and dispersion dominate, plumes contract or attenuate back toward their source. When the plume begins to attenuate, the actual distribution of MTBE in ground water will depend on the relationship between two rates of attenuation: the rate of attenuation of the source and the rate of attenuation along the flow path in the plume. Methods to calculate these rates from field data are presented in Newell et al., (2002).

The long-term source of MTBE at gasoline spill sites is MTBE retained in residual gasoline trapped in the aquifer. The longevity of the plume will depend on the amount of MTBE in the residual gasoline, on the rate that MTBE is transferred from the residual gasoline to ground water, and the rate that the flow of ground water carries the MTBE away from the residual gasoline. Transfer of MTBE will weather the residual gasoline and, over time, will reduce the concentrations in the ground water that are in contact with the residual gasoline. The rate of weathering is the rate of natural attenuation of the source which determines how long a plume will persist over time. Biodegradation, dilution, dispersion and mixing will attenuate concentrations of MTBE as water moves away from the source. They determine the rate of attenuation along the flow path which determines how far a plume will extend away from the source.

When a plume has come to a steady state, the rate of attenuation over time in all the wells should be zero. As the source starts to weather away and the concentrations of contaminant at the source start to attenuate, there will be a corresponding attenuation in all the monitoring wells down gradient of the source. It is important to not confuse attenuation of concentrations over time in monitoring wells down gradient of the source with attenuation along the flow path in the aquifer.

Table 2.1 presents data from six plumes that contrast the roles of these two distinct rates of natural attenuation. The data are from plumes described by Landmeyer et al., (2001) and Wilson and Kolhatkar (2002). All of the plumes in Table 2.1 are old releases that had reached a steady state at the time of the study. The method to calculate the rates of attenuation is described in detail in Newell et al., (2002). The rate of attenuation of the source is calculated from Equation 2.2 as the first order rate of attenuation over time of concentrations of MTBE in the most contaminated well. The rate of attenuation of the plume was calculated from Equation 2.3 as the first order rate of attenuation of concentrations of MTBE with distance along the flow path in the plume, multiplied by an estimate of the plume's seepage velocity. The method to estimate the confidence intervals is illustrated in *Section 7 Statistical Evaluation of Rates of Attenuation of Sources.*

Table 2.1 Rates of Attenuation of MTBE in Source Areas Over Time Contrasted to Rates of Attenuation Along Flow Paths in Ground Water

Location	Attenuation of Plume		Attenuation of Source	
	Rate	Slower 90% Confidence Interval	Rate	Slower 90% Confidence Interval
	per year			
Brandon, FL	2.02	0.71	0.27	0.14
Elizabeth City, NC	1.80	1.20	0.15	0.04
Long Island, NY	0.79	0.53	0.75	0.29
Parsippany, NJ	1.17	0.61	0.19	0.15
Port Hueneme, CA	0.56	0.47	0.23	0.09
Laurel Bay, SC	<0.04	not significant at 90% confidence	0.70	0.60

In the plumes at Brandon, Florida; Elizabeth City, North Carolina; Parsippany, New Jersey; and Port Hueneme, California, the rate of attenuation along the flow path in the plume is faster than the rate of attenuation over time in the source area. As they age, these plumes tend to recede back on themselves. The tendency to recede back to the source is illustrated in Figure 2.3 with data from the plume at Parsippany, New Jersey.

In the plume on Long Island, New York, the source area was remediated. After remediation, the residual "hot spot of MTBE" moved down gradient over time. In the plume at Laurel Bay, South Carolina, the rate of attenuation over time in the source area was faster than attenuation in the ground water. Over time, the "hot spot of MTBE" in the plume detached from the source area and moved down gradient. The tendency for the "hot spot" to detach and move down gradient is illustrated in Figure 2.4 with data from the plume at Laurel Bay, South Carolina.

There was no significant natural biodegradation of MTBE in the anaerobic portion of the plume at Laurel Bay, South Carolina. Because there was little variation in the direction of ground water flow, there was little contribution of dispersion to attenuation. The plume continued to grow until it approached its point of discharge to surface water drainage in a concrete-lined ditch.

As the plume approached the ditch, it was oxygenated as it mixed with ground water in the bed sediments beneath the concrete liner (Landmeyer et al., 2001). On a sampling date in the winter, all of the MTBE in the plume was biodegraded before the plume discharged. On another date in the summer, more than 96% of the MTBE in the plume was biologically degraded before the plume discharged to surface water. The bed sediments of many surface water streams have a considerable capacity for aerobic biodegradation of MTBE (Bradley et al., 2001c).

In general, when the rate of attenuation of the source over time is faster than the rate of attenuation along the flow path, the "hot spot" will detach from the source and move down gradient. This is true whether the attenuation of the source is purely natural, or is a result of the efforts to control or remediate the source area. In general, when the rate of attenuation of the source over time is slower than the rate of attenuation along the flow path, the plume will appear to recede back to the source area over time.

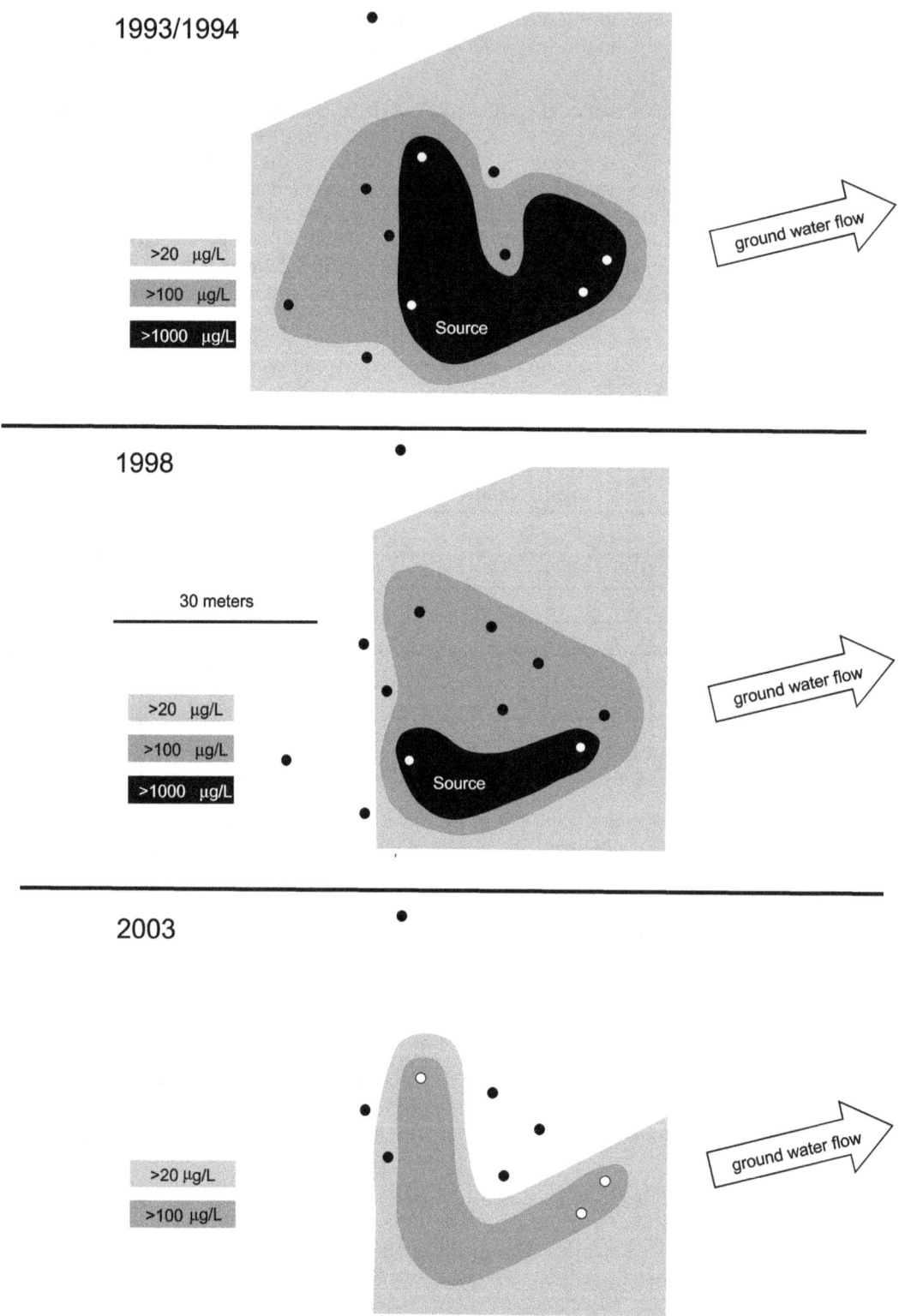

Figure 2.3 *Evolution of a plume of MTBE at Parsippany, New Jersey, that is receding back on itself.*

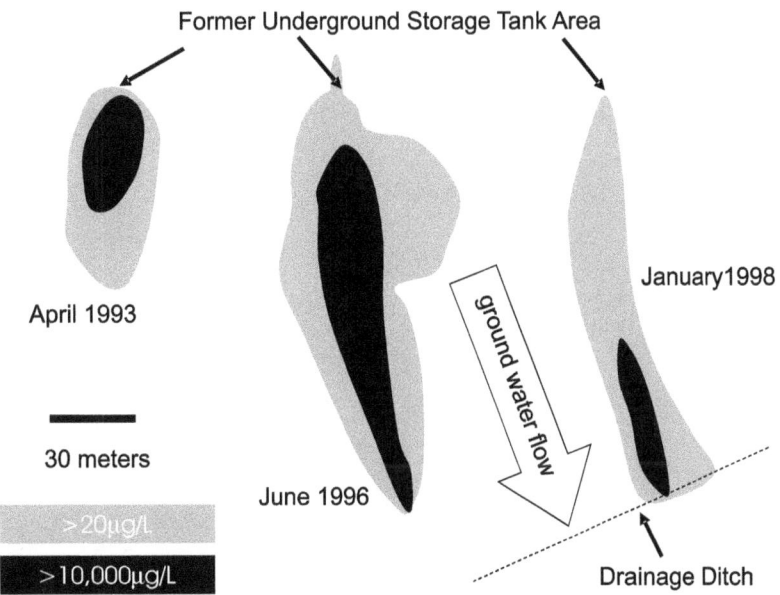

Figure 2.4 *Evolution of a plume of MTBE in Laurel Bay, South Carolina, where the "hot spot" has detached from the source and moved down gradient.*

Section 3 -

Recommendations for Monitoring -

This section discusses a number of interrelated issues concerned with monitoring the behavior of MTBE plumes. Data that are collected to understand the behavior of MTBE at gasoline spill sites are generally collected for three purposes. The primary purpose is to document the distribution of MTBE contamination at the site in time and space. A second purpose is to determine if the coverage of the monitoring wells is adequate to properly determine the distribution of contamination. A third purpose is to predict the physical, chemical, and biological processes that control the distribution of the fuel oxygenates

This section discusses biogeochemical parameters that can be used to recognize the footprint of a plume of contamination from a gasoline spill, even when the concentrations of MTBE have declined below detection limits. This information is valuable to evaluate the coverage of monitoring wells. If biogeochemical data are collected in each round of sampling, they can also be used to determine if a plume has shifted its location.

The distribution of MTBE in ground water is controlled by the direction of ground water flow and by changes in the direction of ground water flow over time. The direction of ground water flow is usually inferred from the elevation of the water table in monitoring wells. Information on the depth to water is usually collected with each round of sampling, but reports to regulators often present ground water contours for only one "typical" round of sampling, or a few rounds of sampling. The flow direction in the "typical" round of sampling becomes ingrained in the site conceptual model; and the variation in the direction of flow is ignored. The variation in flow direction should be considered to determine whether a particular well should sample the plume or sample ambient ground water in each round of sampling. This section illustrates the wide variation in flow direction at a representative field site and introduces two simple computer applications that can be used to extract the direction of ground water flow from data on water elevations in monitoring wells.

The prospects for natural attenuation of chlorinated solvents, or the BTEX compounds, have been correlated with the geochemistry of the ground water. Preliminary studies suggested that anaerobic biodegradation of MTBE is favored under methanogenic conditions and sulfate reducing conditions (Kolhatkar et al., 2000). This section examines the potential correlation in more detail and finds that there is no correlation between the extent of anaerobic biodegradation of MTBE in ground water and the concentration of methane or sulfate in water.

Traditionally, microcosm studies have been used to demonstrate that microorganisms at a site can degrade a contaminant. Microcosm studies of MTBE biodegradation are expensive, time consuming, and often yield equivocal results. As a consequence, they are rarely done as part of the risk evaluation at gasoline spill sites. Either the possibility of natural MTBE biodegradation is ignored altogether, or rate constants published in the literature are extrapolated to a site without any site specific evidence that they are appropriate.

An analysis of the change in the stable carbon isotope ratios in MTBE can provide unequivocal evidence for biodegradation of MTBE at field scale. Unfortunately, at this writing, these analyses are not currently offered by commercial analytical laboratories. They are only commercially available from a few university laboratories. At many sites, the onset of anaerobic biodegradation of MTBE can be recognized by a change in the ratio of TBA to MTBE in the monitoring record. This section illustrates the use of data on the relative concentration of MTBE and TBA as a practical alternative to microcosm studies or stable carbon isotope analyses.

Conventional monitoring wells can provide an incomplete picture of the true distribution of MTBE in ground water. If the screen of a monitoring well is long compared to the thickness of the plume of contamination, it can sample the plume of contaminated ground water and cleaner ground water above or below the plume, giving a false impression of natural attenuation from one well to another. Long plumes of MTBE may dive below the screens of monitoring wells altogether. Any evaluation of natural attenuation between monitoring wells should consider the screened intervals of the wells, the depth interval contaminated with gasoline (if that information is available), and the lithological features sampled by the wells. This report does not further discuss the vertical spacing of monitoring wells in the assessment of natural attenuation and the evaluation of risk. These considerations are discussed in detail in *Performance Monitoring of MNA Remedies for VOCs in Ground Water* (Pope et al., 2004), which is available on the Kerr Center web page. Search for "Kerr" on the U.S. EPA web page.

3.1 Biogeochemical Footprints to Evaluate Coverage of Monitoring Wells

If natural attenuation is reducing the concentrations of contaminants in ground water, the concentrations should be higher in the source area and lower in the down gradient wells. The concentration in monitoring wells down gradient of a source area can be lower because the contaminants were attenuated, or concentrations may be lower because the monitoring well missed the plume (Wilson, 2003a). To distinguish between the two, the National Research Council recommended the use of biogeochemical footprints to distinguish attenuation of contaminants in ground water that has been impacted by a fuel spill from clean water that was never impacted. (NRC, 2000; Rittmann, 2003).

Many organic materials can be metabolized in ground water through an oxidation / reduction reaction where the organic material is oxidized to carbon dioxide while an electron acceptor is reduced. Oxygen, nitrate, and sulfate are often referred to as soluble electron acceptors. During metabolism they are reduced to water, nitrite or molecular nitrogen, and sulfide. Metabolism of the organic matter can be recognized by the depletion of oxygen, nitrate, and sulfate and the accumulation of nitrite or sulfide in water. Iron (III) minerals in sediments can also serve as an electron acceptor. In the process, iron (III) is reduced to iron (II). Although iron (III) minerals are not very soluble in ground water, the iron (II) that is produced from microbial metabolism is more soluble and can accumulate in ground water.

Not all metabolism requires an external electron acceptor. Some organic materials, including the BTEX compounds, can be fermented. One end product of the fermentation of BTEX compounds is methane.

In summary, the metabolism of the BTEX compounds in gasoline can consume oxygen, nitrate, and sulfate or produce methane and iron (II) (Wiedemeier et al., 1995). If changes in these biogeochemical parameters can be associated with a particular spill of gasoline, as distinct from organic materials already present in the aquifer, they can be used as footprints for the plume of contamination produced from the spill.

If the biogeochemical parameters are consistent with the ambient conditions in the aquifer and contaminants are absent, this situation indicates that the well has not been impacted by the gasoline spill. The well is outside the footprint of the plume. If biogeochemical parameters show the depletion of oxygen, nitrate and sulfate, and the accumulation of iron (II) and methane, and the contaminants are absent, this situation indicates that the water was contaminated at one time, but natural attenuation processes have removed the contaminants. The well is inside the footprint of the plume. To be useful as footprints of the plume, these biogeochemical indicator parameters should be measured in each round of sampling where possible.

Biogeochemical footprints are footprints of gasoline contamination as a whole, and not necessarily of MTBE contamination alone. The biogeochemical indicators cannot distinguish a plume of gasoline with MTBE from a plume of gasoline without MTBE. The biogeochemical indicators work best if the MTBE entered ground water from direct contact of gasoline with ground water. If the MTBE entered ground water through a vapor pathway, the readily degradable hydrocarbon components of gasoline may have been removed before the MTBE entered the ground water. They may fail to provide any indication of contamination of ground water by MTBE vapors.

Some of the biogeochemical parameters are more useful than others. Bacterial communities acclimate readily to degrade BTEX compounds using oxygen and nitrate as electron acceptors. Depletion of oxygen and nitrate should be expected at almost every gasoline spill. Bacterial communities also acclimate readily to degrade sulfate. Depletion of sulfate should be expected at most sites as well. Bacterial communities require from months to years to acclimate to ferment BTEX compounds to methane.

The depletion of dissolved oxygen is the most sensitive indication of contamination with gasoline. It is also the most problematic. It is difficult to prevent reoxygenation of ground water samples. It is almost impossible to prevent reoxygenation of the ground water sample if the well is purged and sampled with bailers. Oxygen meters may provide reliable data if they are properly maintained and are recalibrated in the field; but if the appropriate quality assurance procedures are not implemented, they can produce data that are misleading. Simple field test kits can also provide usable data on the concentration of dissolved oxygen in ground water (Wilkin et al., 2001).

Nitrate is the next most sensitive indicator of gasoline contamination. Sampling of nitrate is not problematic, as is the case with oxygen. Analysis of water samples for nitrate is straightforward and inexpensive. Unfortunately, nitrate is often absent under natural conditions in many ground waters. If nitrate is present under ambient conditions, it should be considered as an alternative to dissolved oxygen if the monitoring wells are sampled with bailers.

Sulfate is the most important soluble electron acceptor in ground water at most fuel spill sites (Wiedemeier et al., 1995). Sulfate samples do not require preservation, and analysis for sulfate is straightforward and inexpensive. At most fuel spills, the depletion of sulfate is the best single indicator of impact from a gasoline spill. If it is only possible to monitor one biogeochemical parameter, sulfate should be the parameter of choice.

The accumulation of methane as a tracer works best in old spills. Often the microbial communities do not acclimate to produce methane. Perhaps one third of sites will fail to accumulate methane to concentrations above 0.5 mg/L (Kolhatkar et al., 2000). Methane samples must be collected in a manner that avoids losses due to volatilization, and the samples must be preserved.

The BTEX compounds in gasoline will support iron reduction in almost every site. However, iron (II) often fails to accumulate in the ground water. In many contaminated aquifers, iron reduction and sulfate reduction occur at the same time. The iron (II) will react with any sulfide produced by sulfate reduction and precipitate. If the rate of sulfide production exceeds the rate of iron (II) production, iron (II) may never accumulate in the ground water. Iron is useful as a tracer at less than one-fourth of sites (personal experience of authors). Iron (II) is best determined immediately after water samples are collected.

3.2   Monitoring the Direction of Ground Water Flow

The second approach used to identify the possible footprint of a plume is to measure the hydrological properties of the aquifer receiving the fuel spill, characterize the distribution of the gasoline spill, and then calibrate a computer model such as BIOSCREEN (Newell et al., 1996). For model results to be applicable, the model assumptions must be valid for the site.

Most computer models assume that ground water flows in a uniform direction with a uniform velocity. Any spreading of the plume is attributed to dispersion, and when the models are calibrated to field data, the values for longitudinal and horizontal dispersion are adjusted to match the field data. In some aquifers, the direction and speed of ground water flow are stable, and in these aquifers plumes are usually long and narrow. Often the width of the plume down gradient of the source is no wider than the width of the source area, indicating that transverse and vertical dispersion make a minimal contribution to the distribution of MTBE. If there is little spreading of MTBE to the sides of the plume, a conventional ground water model may provide an accurate forecast of the distribution of contamination.

Other plumes appear to spread laterally as well as longitudinally. This apparent lateral dispersion may be the direct result of changes in the direction of ground water flow. When there is significant variation in the direction and magnitude of ground water flow, conventional models can be misleading about the expected footprint of the plume. What appears to be lateral dispersion is really longitudinal dispersion occurring in different directions. In plumes where there is a wide variation in the direction of ground water flow, simplistic ground water models that assume the ground water moves only one direction at one velocity can be misleading.

Figure 3.1 presents data on the direction and magnitude of ground water flow at an MTBE site at Elizabeth City, North Carolina (Wilson, 2003a). The site is near the Pasquotank River, and the average direction of ground water flow is toward the river; however, the flow at any particular time is sensitive to the stage of the river. The plume was monitored monthly for one year. Figure 3.1 presents predictions to the direction and velocity of ground water flow from the monitoring data in each month.

Regression analysis was used to fit a plane through the elevation of the water table in the monitoring wells (Wilson et al., 2000, Srinivasan, 2004). An arrow is used in the figure to represent the direction and velocity of ground water. The arrows are given different shades to allow them to be distinguished in the figure.

It is apparent that the direction and magnitude of flow vary widely from one month to the next at this site. The standard deviation of the direction of ground water flow over 12 months of sampling, as depicted in Figure 3.1, was 23 degrees. One round of sampling, or even a few rounds of sampling at this site, would not be adequate to define the direction and magnitude of ground water flow. At this site, the contaminant plume occupies the area encompassed by the variation in the direction of ground water flow. This is probably true for all sites where the plume has come to a steady state.

Mace et al., (1997) used a similar approach to calculate the standard deviation of the direction of ground water flow at 132 gasoline stations in Texas (Table 3.1). At each site, the direction of flow was estimated from water table elevations on at least ten separate occasions. The direction of ground water flow at most of the sites in Texas (Figure 3.2) was more variable than the site in North Carolina illustrated in Figure 3.1. The median of the standard deviation of the direction of ground water flow in Texas was 36 degrees.

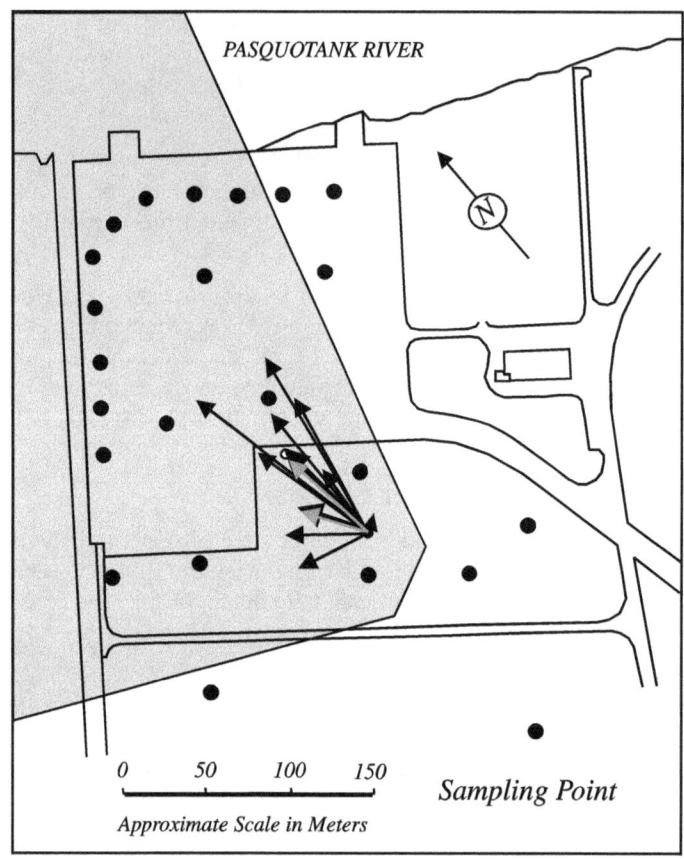

Figure 3.1. *Variation in the direction and magnitude of ground water flow at an MTBE site in Elizabeth City, North Carolina. The arrows represent the distance that water would move in one year, based on the direction and hydraulic gradient present in a particular round of sampling. Some of the arrows are shaded grey to allow them to be distinguished from the other arrows. The origin of the arrows is the center of the LNAPL source area. The black dots are locations of monitoring wells. The shaded area includes all the monitoring wells with concentrations of MTBE above 20 µg/l.*

To facilitate the visualization of the variability of the flow direction of the sites in Texas, Figure 3.2 summarizes the data in Table 3.1 by comparing the arcs subtended by one standard deviation in flow direction on either side of the mean direction of flow. A comparison is made between the sites with low variation (± 20 degrees), moderate variation (± 30 degrees), high variation (± 50 degrees), very high variation (± 70 degrees), and extremely high variation (± 120 degrees). The variability in the direction of ground water flow in Texas is probably typical of many regions of the United States. For roughly one-third of the sites in Texas, the direction in ground water flow is highly variable, and the concept of a single flow direction is not the best representation of the behavior of the plume. A conventional ground water model could be misleading at these sites.

Site investigation reports may include maps showing the contour of the water table, particularly if there are significant variations in the direction of flow from one sampling event to the next. However, the variation in the direction of ground water flow is rarely evaluated in any formal way. As a consequence, the monitoring wells on the perimeter of a site may not be in the best position to detect a plume of contamination.

Data on water table elevations are frequently collected at gasoline spill sites. The U.S. EPA has created a decision support tool that evaluates data on water table elevations in monitoring wells to make predictions of the most likely footprint of a plume of contamination. The Optimal Well Locator (OWL) is a screening tool to evaluate the locations of existing monitoring wells, to identify the best locations for new wells, and to identify the existing wells that are least likely to detect the plume (Srinivasan, 2004). OWL was used to generate the predictions on the direction and magnitude of ground water flow that are presented in Figure 3.1. A simple calculator for the direction of ground water flow is also available on the Athens Laboratory web page. (http://www.epa.gov/athens/learn2model/part-two/onsite/index.html)

Table 3.1 - Variation in the Standard Deviation of the Direction of Ground Water Flow at 132 Gasoline Stations in Texas (Data from Figure 14 in Mace et al., 1997)

Standard Deviation (degrees)	Occurrence	Frequency %	Cumulative Frequency %
0 to 10	6	4.6	4.6
10 to 20	22	16.8	21.4
20 to 30	22	16.8	38.2
30 to 40	23	17.6	55.7
40 to 50	10	7.6	63.4
50 to 60	14	10.7	74.0
60 to 70	7	5.3	79.4
70 to 80	9	6.9	86.3
80 to 90	13	9.9	96.2
90 to 100	3	2.3	98.5
100 to 110	1	0.8	99.2
110 to 120	1	0.8	100.0

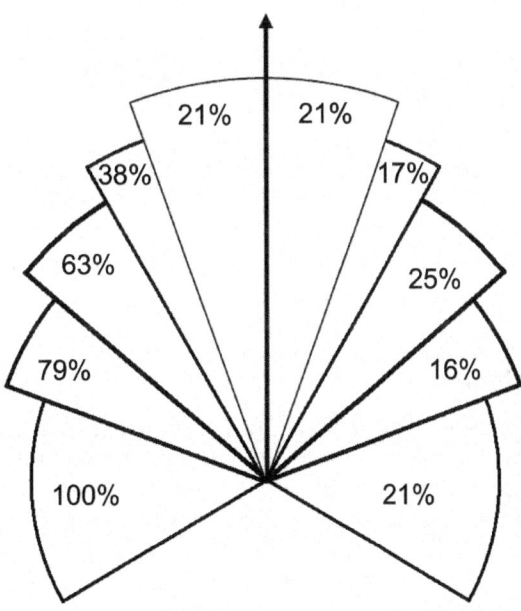

Figure 3.2. - *Variation in the direction of ground water flow at gasoline spill sites in Texas. The solid arrow represents the direction of ground water flow. Presented are the arcs subtended by one standard deviation of the direction of ground water flow. Numbers on the left side of the figures are the percentage of 132 sites where one standard deviation is contained within the arc to the left of the number. Numbers on the right hand side of the figure are the percentage of 132 sites where the arc subtended is greater than the arc to the left of the number but less than the arc to the right of the number in the figure.*

3.3 Monitoring to Predict Biological Processes

If the biogeochemical parameters are used to determine the chemical environment of the organisms that degrade MTBE in the aquifer, it is not necessary to determine oxygen, nitrate, sulfate, methane, and iron (II) at every round of sampling. Oxygen should be monitored as a routine parameter only when the monitoring wells are sampled using pumps. If the wells are sampled with bailers, it is best not to attempt to measure dissolved oxygen in the ground water. If the concentration of oxygen in the well water is less than 0.5 mg/L, there is not an adequate supply of oxygen to support extensive metabolism of MTBE.

At gasoline spill sites, the vertical gradients of biogeochemical parameters may be strong. It is not unusual for a monitoring well that is screened across the water table to produce oxygenated uncontaminated water from the depth interval right at the water table and water that is contaminated and strongly anaerobic from an interval less than a meter deeper. The best indication of anaerobic conditions in the contaminated portion of the aquifer is the presence of iron (II) or methane in water from the monitoring well. The blended water produced by the well may contain oxygen even when oxygen is totally absent in the contaminated portion of the aquifer.

Analysis of methane in samples can be expensive and is not recommended for routine monitoring. The measured concentrations of iron (II) in ground water have no straightforward relationship to the amount of biological activity in the water. The presence of iron (II) in water at detectable concentrations indicates that biological iron reduction is occurring in the aquifer. The actual concentration of iron (II) has no direct interpretation. Routine monitoring for iron (II) is not recommended.

Early work on natural anaerobic biodegradation of MTBE indicated an association of MTBE biodegradation with methanogenic conditions in ground water (Wilson et al., 2000; Kolhatkar et al., 2000; Kolhatkar et al., 2001; Kolhatkar et al., 2002). More recent work at thirteen gasoline spill sites in Orange County, California, used the stable carbon isotope ratio of MTBE in ground water to estimate the extent of natural biodegradation (Kuder et al., 2005; Wilson et al., 2005b, discussed in detail in Sections 5 and 6).

Figure 3.3 compares the extent of biodegradation of MTBE in ground water at the thirteen sites in Orange County, California, to the concentration of methane or sulfate in the water. There was no clear association of natural MTBE biodegradation with high concentrations of methane or with low concentrations of sulfate in the ground water (Figure 3.3). Although these biogeochemical parameters may be useful to recognize the footprint of the plume from a spill of gasoline, these parameters have little value to predict anaerobic biodegradation of MTBE.

The first biodegradation product of MTBE is TBA. At many sites, the TBA produced from MTBE biodegradation accumulates in the ground water. If the long-term monitoring record includes analysis of TBA, many sites show a transition in the relative proportions of TBA and MTBE in the ground water (see Figure 3.4 for an example). The best indication for MTBE biodegradation that is available from conventional monitoring parameters is an abrupt and persistent increase in the ratio of TBA to MTBE in ground water.

Depending on its concentration, the TBA produced by natural biodegradation of MTBE in the ground water may also be a concern.

3.4 Concerns with Analytical Issues

Good environmental monitoring requires good chemistry. The conventional analytical approach for analysis of gasoline components in ground water is preparation with a purge-and-trap unit, separation on a gas chromatograph, and determination with a flame ionization detector (EPA method 8015). Crumbling and Lesnik (2000) noted that alcohols, such as TBA, are not efficiently recovered from water by purge-and-trap, and as a result, the analytical detection limits for TBA and other fuel alcohols are high. Often the reporting limit is above the Provisional Action Goal set by California for TBA of 12 µg/L. To improve the recovery efficiency of TBA and other fuel alcohols, Lin et al., (2003) prepared water samples using a heated headspace sampler. With a heated headspace sampler and determination with a mass spectrometer (EPA method 8260), they could achieve a method detection limit of 0.8 µg/L for TBA.

MTBE is manufactured from a mixture of isobutylene and methanol, with the aid of an acid (H^+) catalyst. Acid can also catalyze the hydrolysis of MTBE to produce TBA and methanol. If ground water is preserved with acid to pH<2, MTBE can be hydrolyzed to TBA (Lin et al., 2003; McLoughlin et al., 2004; O'Reilly et al., 2001; White et al., 2002). In samples that are refrigerated, the rate of hydrolysis is slow. When samples are stored at 10° C, less than 5% of the MTBE is hydrolyzed within 30 days. There is little hydrolysis during sample preparation by purge-and-trap at room temperature. However, when the water sample is heated to improve the recovery of TBA, a major fraction of the MTBE is hydrolyzed during analysis. Hydrolysis of MTBE to TBA can be avoided by preserving the sample with 0.1% trisodium phosphate instead of acid (Lin et al., 2003; McLoughlin et al., 2004; White et al., 2002).

As will be discussed in detail in Sections 5 and 6, the most unequivocal indicator of natural MTBE biodegradation is the ratio of stable carbon isotopes in the residual MTBE. Samples collected for analysis of stable carbon isotope ratios should be preserved with trisodium phosphate whenever possible. The effective holding time for samples preserved with trisodium phosphate is more than three months. If the analytical laboratory is backed up and the samples cannot be analyzed within a few weeks, the samples will not be compromised by the longer holding times.

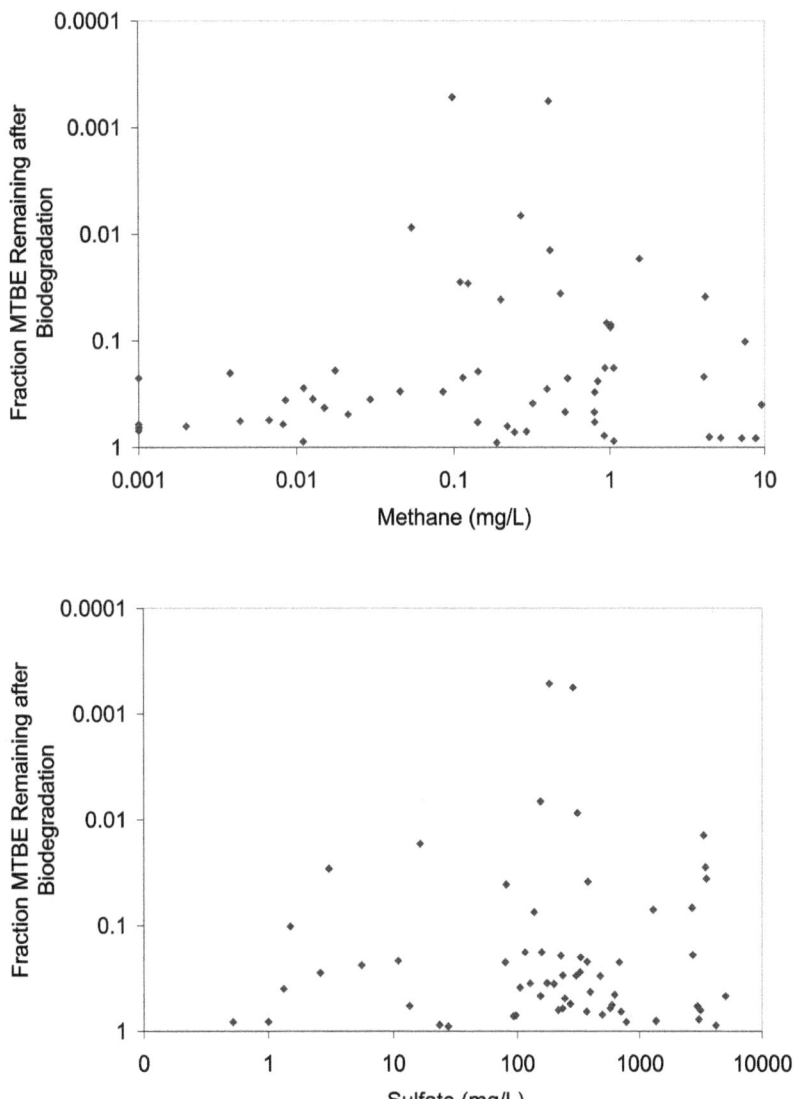

Figure 3.3. *Association of sulfate and methane with natural anaerobic biodegradation of MTBE in ground water collected from 61 monitoring wells at thirteen gasoline spill sites in Orange County, California.*

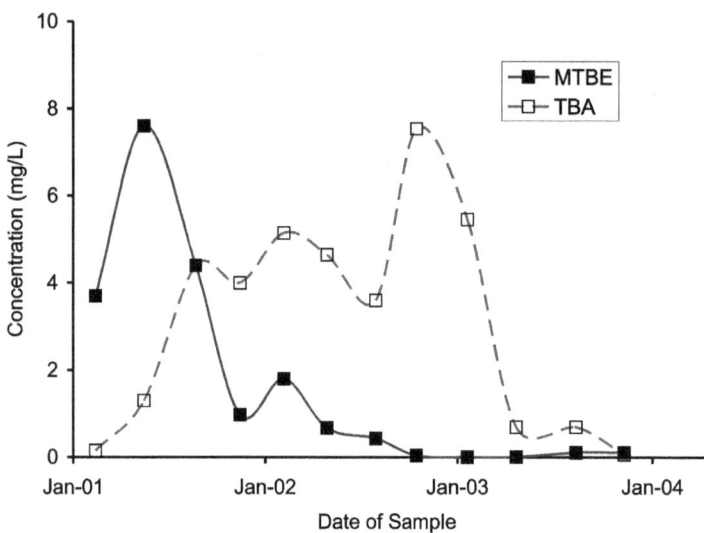

Figure 3.4. *A change in the relative concentration of TBA and MTBE in a monitoring well as evidence for natural biodegradation of MTBE. Data are from a gasoline spill site in Tustin, California.*

Section 4 -

Biological Degradation -

Many ground water scientists and engineers think that MTBE does not biodegrade in ground water. The idea may come, in part, from the fact that it is difficult to culture MTBE degrading bacteria. The idea may also come, in part, from the fact that many MTBE plumes are much longer than the associated plumes of benzene and BTEX compounds that come from the same spill of gasoline. Finally, the idea that MTBE does not degrade is consistent with the behavior of MTBE in a controlled release experiment that was done by the University of Waterloo at Canadian Forces Base Borden in Ontario, Canada. In an aquifer that had not experienced MTBE contamination, MTBE did not degrade in the first years of the experiment. In subsequent years, MTBE did biodegrade in the field-scale plume, but by then, the idea that MTBE does not biodegrade was fixed in the minds of many ground water professionals.

This section briefly reviews the current state of knowledge concerning microbiology of MTBE biodegradation. The information provided in this section draws from information provided in several recent reviews of the biodegradation of MTBE (Deeb et al., 2000a; Cozzarelli and Baehr, 2003; Fiorenza and Rifai, 2003; Finneran and Lovley, 2003; Rittmann, 2003; Wilson, 2003a; Wilson, 2003b, and Schmidt, 2004). The review by Schmidt et al., (2004) is particularly detailed and comprehensive.

There are three general approaches to document biodegradation; the loss of the parent substrate, the accumulation of an intermediate biodegradation product, and mineralization of the organic carbon originally present in the parent substrate. Mineralization studies are conventionally done by labeling the parent compound with ^{14}C, and measuring the accumulation of radio-label in carbon dioxide or methane. All three approaches have been applied to understand biodegradation of MTBE.

4.1 Microbiology of Aerobic MTBE Biodegradation

Organisms that grow aerobically on MTBE are difficult to isolate and culture in the laboratory. In laboratory microcosms, Yeh and Novak (1994) found no evidence for aerobic biodegradation of MTBE after 100 days of incubation. After 60 days of incubation, Jensen and Arvin (1990) found no degradation of MTBE in samples of activated sludge, topsoil, or aquifer material. In 1994, Salanitro et al. published the first report of the aerobic biodegradation of MTBE by a mixed culture of microorganisms; in 1997, Mo et al. published the first report of aerobic biodegradation of MTBE by pure cultures of bacteria; and in 1997, Borden et al. published the first report of aerobic MTBE biodegradation in microcosms. In the study of Borden et al. (1997), the aerobic biodegradation of MTBE in the microcosms was incomplete; the biodegradation of MTBE stopped after 100 days of incubation, leaving approximately 1.0 mg/L of MTBE in the pore water.

There are several explanations of why it is difficult to isolate MTBE degrading micro-organisms. All of the carbon-to-carbon bonds in MTBE involve bonds with the central tertiary carbon. It is difficult for microorganisms to break a carbon bond with a tertiary carbon atom (one more carbon-to-carbon bond would result in the bonding structure of diamond). It is also difficult for microorganisms to degrade ethers, and MTBE is an ether. If the ether bond is broken by enzymatic hydrolysis, the products are TBA and formaldehyde, and formaldehyde is toxic. Finally, it may be difficult to isolate pure cultures of MTBE degrading organisms because MTBE is degraded in nature by a consortium of different types of microorganisms acting together, and not by single organisms.

Because it was difficult to isolate and culture microorganisms that aerobically degraded MTBE, there was a limited appreciation of the capacity of aerobic bacteria to degrade MTBE in ground water. In their comprehensive review of the behavior of MTBE in the environment, Squillace et al. (1997) concluded, "In general, most studies to date have indicated that MTBE is difficult to biodegrade, and some have classified MTBE as recalcitrant."

This initial difficulty in isolating and culturing MTBE degrading bacteria was caused by the impatience of the scientists as much as the metabolic capability of the microbes. Traditionally, microbiologists working in the laboratory base their expectations, and design their experimental protocols, on the behavior of organisms that grow rapidly. They rarely incubated their enrichment studies for more than a few months.

The difficulty in isolating microorganisms that can degrade MTBE is easily understood if we compare the growth rate of microorganisms that grow on MTBE to the growth rate of microorganisms that degrade ordinary petroleum hydrocarbons such as benzene, toluene, and xylenes (Table 4.1). Typical strains of bacteria growing aerobically on petroleum hydrocarbons can divide and double their numbers every two to five hours at room temperature. As a consequence, laboratory enrichment cultures will grow up and remove the hydrocarbons in a few days. On the other hand, cultures of bacteria using MTBE as a growth substrate require several days to several weeks to double their numbers. Their growth rate is from one-tenth to one-hundredth of the growth rate of bacteria that degrade conventional petroleum hydrocarbons. Their very slow growth rate has an important effect on the time required for a culture to grow to densities that will entirely consume MTBE.

Table 4.1. Comparison of the Growth Rate of Aerobic Bacteria During Growth on MTBE as the Primary Substrate to the Growth Rate of Bacteria that Grow on Pentane and Fortuitously Metabolizes MTBE and to the Growth Rate of Bacteria that Grow on BTX Compounds

Growth Substrate	Source of Data	Doubling Time (hours)	Reference
Organisms that can degrade MTBE as the primary substrate			
MTBE	BC-1 culture	>340	Salanitro et al. (1998)
MTBE	Enrichment from Refinery Activated Sludge	58	Park and Cowan (1997)
MTBE	Enrichment from Biofilter	670	Fortin and Deshusses (1999)
MTBE	ENV735 *Hydrogenophaga flava*	41	Steffan et al. (2000)
Organisms that co-metabolize MTBE			
Pentane	*Pseudomonas aeruginosa*	3.6	Garnier et al. (1999)
Organisms that cannot degrade MTBE			
Benzene	Median of 10 studies	2.1	Suarez and Rifai (1999)
Toluene	Median of 15 studies	2.9	Suarez and Rifai (1999)
Xylenes	Median of 8 studies	5.0	Suarez and Rifai (1999)

The BC-1 culture acquired by Salanitro et al., (1998) requires at least 340 hours to double its density (see Table 4.1). The complete metabolism of an initial concentration of MTBE of 1.0 mg/L will produce a final density of bacteria of approximately 10^6 per milliliter. If the initial density of the BC-1 culture were one active cell per milliliter, it would require 20 cycles of cell division over 283 days for the culture to grow up and degrade MTBE. Strain ENV735 is another bacterium that can degrade MTBE as the sole carbon source. Its doubling time is near 41 hours; it would require 34 days for the culture to grow up and degrade MTBE.

Fortin and Deshusses (1999) note that most of the cultures of MTBE degrading bacteria that were available to them at that time were acquired from bioreactors or biofilters that had already acclimated to degrade MTBE. It is likely that early attempts to enrich for the organisms that degrade MTBE failed because the enrichment cultures were not incubated for an adequate period of time.

Some organisms can biodegrade MTBE, but they cannot grow on MTBE alone; these organisms require another substrate for growth. Biodegradation of MTBE under these circumstances is termed co-metabolism or co-oxidation. The organisms that grow on other substrates and co-metabolize MTBE can grow rapidly. Compare the growth rate of *Pseudomonas aeruginosa* when growing on pentane to the growth rate of microorganisms growing on MTBE (Table 4.1). With a doubling time of 3.6 hours, this organism could grow from an initial density of one cell per ml up to densities that can degrade 1.0 mg/L of MTBE in only three days.

Many of the natural hydrocarbons in gasoline can support the growth of organisms that will degrade MTBE. This is particularly true of the straight-chained alkanes and iso-alkanes (Hyman et al., 2000). Because they grow more rapidly, adding oxygen to environmental samples that contain a mixture of petroleum hydrocarbons and MTBE will most likely enrich for organisms that co-metabolize MTBE.

4.2 Biochemistry of Aerobic MTBE Biodegradation

A simplified and generalized pathway for complete aerobic metabolism of MTBE is presented in Figure 4.1. The figure combines features in the pathways published by Fiorenza and Rifai (2003), Steffan et al., (1997) and Deeb et al., (2000a). In every aerobic organism studied to date, the first transformation is believed to be carried out by a mono-oxygenase enzyme. These enzymes insert one oxygen atom from molecular oxygen into the organic compound being metabolized. The other oxygen atom is reduced to form water. The first stable products are TBA and either formaldehyde or formic acid. Formaldehyde and formic acid are very readily degraded.

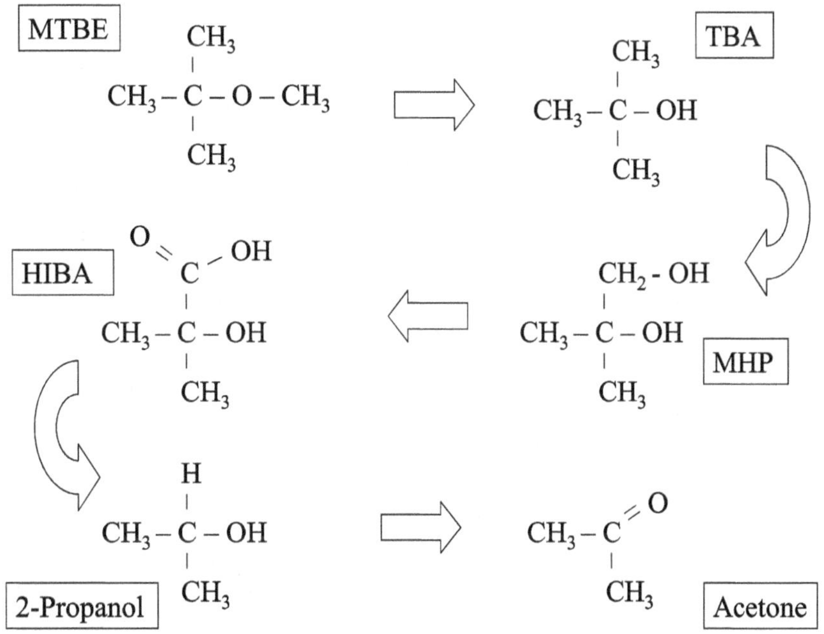

Figure 4.1. *Significant products of the aerobic biodegradation of MTBE. (After Wilson 2003b). MHP is methyl-2-hydroxy-1-propanol. HIBA is 2- hydroxyisobutyric acid.*

Often the resulting TBA will accumulate in ground water. Kane et al., (2001) showed the transitory accumulation of TBA during aerobic biodegradation of MTBE by naturally occurring bacteria from a gasoline spill site in Palo Alto, California. Hunkeler et al. (2001) showed transitory accumulation of TBA in laboratory cultures during aerobic growth on MTBE and during aerobic co-metabolism of MTBE supported by 3-methylpentane. Apparently, the native microbial population at the spill site included organisms that could degrade MTBE and TBA. The MTBE was degraded first, and then the TBA was degraded if the supply of oxygen was sufficient.

The TBA can be further transformed through a second attack by a mono-oxygenase to form 2-methyl-2-hydroxy-1-propanol (MHP). The MHP is further oxidized to 2-hydroxyisobutyric acid (HIBA). HIBA has been detected in ground water at a gasoline spill site (Personal Communication, Pat McLoughlin, Microseeps Inc., Pittsburg, PA). Elimination of the carboxylic acid group from HIBA produces 2-propanol (isopropyl alcohol), which in turn can be oxidized to acetone.

Acetone and 2-propanol are rapidly degraded in aerobic ground waters to carbon dioxide, water, and biomass. These compounds should be more persistent in anaerobic ground water than in aerobic ground water, but they should eventually degrade. Acetone is occasionally reported in ground water from gasoline spills. It has conventionally been attributed to contamination of the field sample by acetone in the laboratory. There is a strong possibility that the acetone reported in ground water samples from gasoline spills was a biodegradation product of MTBE or TBA.

4.3 Aerobic MTBE Biodegradation in Ground Water

In 1988, the University of Waterloo conducted a large controlled-release study of MTBE degradation in a sandy glacial aquifer at Canadian Forces Base Borden in Ontario, Canada, (Hubbard et al., 1994; Schirmer and Barker 1998; Schirmer et al., 1999). They injected ground water containing 19 mg/L BTEX and 269 mg/L MTBE into the aquifer and monitored the degradation of BTEX and MTBE in the plume. The BTEX compounds were completely removed within 476 days. However, there was no statistically significant evidence for biodegradation of MTBE (Hubbard et al., 1994). Many of the readers of Hubbard et al. (1994) interpreted the lack of evidence for biodegradation of MTBE as evidence that MTBE would not biologically degrade in ground water. This report supported and reinforced the conventional wisdom at the time that MTBE did not biodegrade in aquifers.

Researchers at the University of Waterloo sampled the MTBE plume again in 1995. The concentrations of MTBE were much lower than expected based on dilution and dispersion alone. In 1996, they sampled the plume using a fine grid to give themselves greater confidence in their estimate of the mass of MTBE remaining in the aquifer. After 3,000 days of residence time, only 3% of the MTBE originally injected into the aquifer remained in the aquifer (Schirmer and Barker, 1998; Schirmer et al., 1999). When they used sediment from the aquifer to construct laboratory microcosms, acclimation to degrade MTBE was a rare event; only 3 of 40 microcosms acclimated after 20 months of incubation. However, once the acclimation event occurred in laboratory microcosms, biodegradation was rapid and extensive. Based on this finding, they attributed the disappearance of MTBE in the field scale plume to aerobic biodegradation.

There seems to be a wide variation from one gasoline spill to another in the distribution and activity of native microorganisms that can degrade MTBE. Salinitro et al., (1998) surveyed sites for the presence of MTBE degrading bacteria. They examined ground water and soil from ten sites: two retail sites in California, refineries in Louisiana and Illinois, distribution terminals in Nevada and Ohio, a pipeline in Texas, and retail sites in Michigan, Texas, and New Jersey. They were able to isolate MTBE degrading organisms from two of the ten sites and demonstrate MTBE degradation in microcosms constructed with material from two sites. Similarly, Kane et al., (2003) [see also Kane et al., (2001)] constructed microcosms with material from seven MTBE spills in California; MTBE was degraded in sediment from only three of the sites. In sediment from the other four sites, MTBE was not degraded within the period of incubation (170 days to 350 days depending on the site).

Hanson et al., (1999) reported the isolation of strain PM-1, a pure culture that can degrade MTBE as the primary substrate. The strain was isolated from a biofilter that was used to treat the off-gases from a sewage treatment plant that received discharges from a local refinery. Recent developments in molecular genetics make it possible to identify the genes of particular bacteria in samples of aquifer material. Using denaturing gradient gel electrophoresis, Kane et al., (2003) showed that the sediment from all three of the gasoline spill sites in California that degraded MTBE harbored bacteria with DNA similar to strain PM-1. Sediment that did not degrade MTBE did not harbor bacteria with DNA similar to PM-1. Hristova et al., (2003) isolated bacteria containing DNA very similar to PM-1 from an aerobic biological treatment system for MTBE at a gasoline spill site on Vandenberg AFB, California. Many of the bacteria that have been studied to date that degrade MTBE under aerobic conditions in ground water seem to be closely related to strain PM-1.

Strain PM-1, which can grow on MTBE as a sole carbon source, also grows readily on benzene. While PM-1 is growing on benzene, it does not degrade MTBE. Once the benzene is exhausted, it produces the enzymes necessary to degrade MTBE and starts to grow on MTBE (Deeb et al., 2000a; Deeb et al., 2000b; Deeb et al., 2001). Benzene present in a gasoline spill could enrich PM-1 to high density and prepare the site for rapid MTBE biodegradation once the BTEX compounds were exhausted. Hyman (2000) noted that organisms that can degrade alkanes and isoalkanes and can cometabolize MTBE are common in ground water at gasoline spill sites. Alkanes, isoalkanes, and the BTEX compounds are the major components of gasoline. These components of gasoline may enrich for organisms that can degrade MTBE once the gasoline has been exhausted.

The prospects for natural aerobic biodegradation of MTBE by native microorganisms may be related to the age of the spill, the time that has been available for acclimation of the native microorganisms to MTBE, and perhaps to the seepage velocity of ground water. Because the organisms that can degrade MTBE grow so slowly, acclimation to degrade MTBE may require several years, as was the case at Canadian Forces Base Borden in Ontario, Canada. If a release starts as a slow pinhole release, and only later grows large enough to be noticed, there may be time for acclimation before the major portion of the release occurs. When oxygen is added to support aerobic biodegradation of MTBE in old anaerobic plumes, they often acclimate in weeks to months (Salanitro et al., 2000; Wilson et al., 2002).

Can the leading edge of an MTBE plume outrun the bacteria and escape biodegradation? Not in the long term. Although most bacteria in aquifers are associated with surfaces, many of them are planktonic. The planktonic bacteria are already in the ground water and move with the ground water. Any transport process that will advance the MTBE will advance the bacteria. Their motion is with respect to the plume itself. They move within the plume even as the plume advances through the aquifer. In general, flagellated planktonic bacteria can move no more than 6 cm a day through flowing ground water. If the seepage velocity of a plume is high, a plume may get to be very large before the acclimation event occurs. Once acclimation has occurred at a particular point, it may take a long period of time for the bacteria to spread throughout the rest of the plume.

4.4 Anaerobic Biodegradation of MTBE

There are reports in the literature of MTBE biodegradation under nitrate-reducing conditions (Bradley et al., 2001a), sulfate-reducing conditions (Bradley et al., 2001b; Somsamak et al., 2001), iron-reducing conditions (Landmeyer et al., 1998; Bradley et al., 2001b; Finneran, and Lovley, 2003), and methanogenic conditions (Mormile et al., 1994; Wilson et al., 2000; Bradley et al., 2001b; Kolhatkar et al., 2002; Somsamak et al., 2005; Wilson et al., 2005a).

Bradley and co-workers added radio-labeled MTBE to stream bed sediments and compared the distribution of biodegradation products under nitrate-reducing, sulfate-reducing, iron-reducing, and methanogenic conditions (Bradley et al., 2001a and 2001b). Table 4.2 summarizes some of their results. Under nitrate-reducing conditions, the MTBE that was degraded was completely metabolized to carbon dioxide. Under sulfate-reducing conditions, iron-reducing conditions, and methanogenic conditions, TBA accumulated to a greater or lesser extent in the different sediments.

In these experiments, the MTBE was uniformly labeled with ^{14}C. Only 20% of the radio-label in the MTBE added to the sediment was associated with the methoxyl-carbon of MTBE. If the label recovered as carbon dioxide exceeded 25% of the label recovered as TBA, then some portion of the TBA produced from MTBE biodegradation must have been further metabolized. A sulfate reducing culture described by Somsamak et al., (2001) did not degrade TBA. However, in the microcosm study of Bradley et al. (2001b), some portion of the TBA produced during MTBE biodegradation was further oxidized to carbon dioxide under sulfate-reducing conditions in all three sediments tested. In one of the sediments tested by Bradley et al. (2001a, 2001b), most of the TBA produced during biodegradation of MTBE under iron-reducing conditions was further oxidized to carbon dioxide (Table 4.2).

Table 4.2 Distribution (in percent) of Biodegradation Products of MTBE under Nitrate-reducing, Sulfate-reducing, Iron-reducing, and Methanogenic Conditions (Bradley et al., 2001a and 2001b)

Amendment	Location	Duration	MTBE	TBA	CO_2	Methane	Total
		days	Percent of original radio-label in MTBE				
	FL	166	29 ± 2	nd	75 ± 1	nd	104 ± 12
	SC	166	72 ± 1	nd	33 ± 8	nd	105 ± 7
Nitrate	NJ	166	81 ± 2	nd	23 ± 5	nd	104 ± 4
	SC	77	70 ± 1	nd	26 ± 10	nd	96 ± 10
	FL	77	71 ± 4	1 ± 1	23 ± 5	nd	95 ± 5
	FL	166	82 ± 3	1 ± 1	20 ± 4	nd	103 ± 4
Sulfate	SC	166	81 ± 9	9 ± 7	9 ± 3	nd	104 ± 12
	NJ	166	82 ± 3	3 ± 0	12 ± 3	nd	97 ± 2
	FL	166	88 ± 3	9 ± 1	nd	3 ± 2	100 ± 3
Iron (III)	SC	166	92 ± 12	8 ± 4	nd	3 ± 2	102 ± 10
	NJ	166	81 ± 10	4 ± 1	14 ± 4	nd	99 ± 10
	FL	166	82 ± 3	1 ± 1	20 ± 4	9 ± 3	103 ± 4
None, all	SC	166	81 ± 9	9 ± 7	9 ± 3	9 ± 3	104 ± 2
sediments are	NJ	166	82 ± 3	3 ± 0	12 ± 3	9 ± 3	97 ± 2
methanogenic	SC	77	85 ± 5	10 ± 2	3 ± 3	1 ± 1	99 ± 5
	FL	77	78 ± 4	9 ± 2	5 ± 2	5 ± 1	97 ± 4

nd - *means not detected.*

These studies were done with mixed microbial communities in sediments. The redox potential and the exposure to MTBE varied from sediment to sediment. An observation that MTBE biodegradation occurred under methanogenic conditions in the sediment does not mean that the MTBE was degraded by methanogenic organisms. Similarly, an observation that MTBE degradation occurred under iron-reducing conditions does not mean that the MTBE was degraded by iron-reducing organisms. Several electron-accepting processes can occur concomitantly in aquifer material. In a recent review, Schmidt et al., (2004) noted that most of the laboratory studies conducted to date have failed to associate anaerobic MTBE biodegradation with a specific electron accepting process. However, these studies do show that MTBE can be degraded under oxidation-reduction conditions that are common in contaminated ground water at gasoline spill sites.

Somsamak et al., (2005) reported degradation of MTBE to TBA in enrichments from a microcosm that was originally methanogenic. Oxygen, nitrate, sulfate, and iron (III) were not available. When methanogenesis was inhibited with 20 mM 2-bromomethanesulfonic acid (BES), MTBE continued to degrade. Somsamak et al., (2005) speculated that anaerobic MTBE biodegradation in their culture may have been carried out by homoactogenic bacteria. These bacteria can metabolize a methyl-ether using molecular hydrogen and bicarbonate ion to produce acetate and the corresponding alcohol. Wilson et al., (2005b) compared the Gibbs free energy for the metabolism of molecular hydrogen by methanogens to metabolism by homoactogens using MTBE. At concentrations of molecular hydrogen that would be expected in ground water, there is more energy available to the homoactogens, and they would be expected to have a competitive advantage over the methanogens that use hydrogen.

Although these studies prove that anaerobic biodegradation of MTBE in sediments is possible, they do not indicate that anaerobic biodegradation in aquifer sediments is common or pervasive. Amerson and Johnson (2002) added MTBE labeled with ^{13}C to a large MTBE plume at Port Hueneme, California. They found no evidence for loss of the MTBE labeled with ^{13}C over the course of one year. Landmeyer et al. (1998) documented degradation of MTBE under iron-reducing conditions in microcosms constructed with sediment impacted by a gasoline spill site in South Carolina. Although the removal was statistically significant, the rate was very slow (2% ± 0.6% oxidized to CO_2 in four months). The overall rate of biodegradation in the plume in South Carolina was less than 0.04 per year (Landmeyer et al., 2001). Biodegradation would have had minimal influence on distribution of contamination in the plume. Anaerobic biodegradation of MTBE may have little effect on the distribution of the plume of MTBE at many gasoline spill sites.

Wilson and coworkers constructed microcosms with contaminated sediment from aquifers that had been impacted by spills of gasoline or neat MTBE (Wilson et al., 2000; Kolhatkar et al., 2002; Kuder et al., 2002; Kuder et al., 2003; Wilson et al., 2005a; Adair et al., unpublished). Sediment from gasoline spill sites at Boca Raton, Florida; Parsippany, New Jersey; Deer Park, New York; Petaluma, California; and Vandenberg AFB, California; were amended with either 2 mg/L MTBE, or 2 mg/L TBA, or 2 mg/L benzene, or 2,000 mg/L ethanol. Sediment from a gasoline spill site at Port Hueneme, California, was amended with 10 mg/L MTBE or 10 mg/L TBA, but not with benzene. Sediment from a JP-4 jet spill at Elizabeth City, North Carolina, was amended with MTBE or TBA or benzene, but not with ethanol. Sediment from a spill of neat MTBE at a tank farm in Nederland, Texas, was mixed together until the concentration of MTBE and TBA already present in the sediment was uniform, but the sediment was not amended with additional MTBE, TBA, benzene, or ethanol. The sediments were incubated for up to 24 months in an anaerobic glove box. If more than 90% of the material was removed compared to a sterilized control, the material was considered to have biologically degraded.

Benzene degraded under anaerobic conditions in all six of the sediments tested (Table 4.3). The sediments were depleted of oxygen, nitrate, and sulfate before the microcosms were constructed. Biodegradation of benzene in the microcosms occurred under methanogenic conditions or possibly under iron-reducing conditions. Biodegradation of benzene was not tested in sediment from the Port Hueneme, California, site. At this site, benzene degrades readily under sulfate reducing conditions in the field scale plume.

There was no evidence that MTBE degraded under anaerobic conditions at the Port Hueneme, California, site (Amerson and Johnson; 2002), and the Vandenberg AFB, California, site (Wilson et al., 2002). Because MTBE did not degrade at field scale, degradation was not expected in the microcosms. As expected, MTBE did not degrade in sediment from these sites (Table 4.3). There was evidence at field scale that MTBE was degrading under anaerobic conditions at the Parsippany, New Jersey, site (Kolhatkar et al., 2002), at the Elizabeth City, North Carolina site (Wilson et al., 2000), at the Deer Park, New York, site (Kolhatkar et al., 2002), at the Boca Raton, Florida, site (case files), the Petaluma, California, site (case files), and at the Nederland, Texas, site (case files). Because MTBE appeared to degrade at field scale, MTBE degradation was expected in the microcosms. In contrast to the behavior of benzene, MTBE degraded in

Table 4.3 Anaerobic Biodegradation of MTBE, TBA, Benzene, and Ethanol in Microcosms Constructed with Aquifer Sediment

Location	MTBE in plume	TBA in plume	Benzene in plume	MTBE degraded	TBA degraded	Benzene degraded	Ethanol degraded
	µg/L						
Boca Raton, FL	106	11,600	132	No	No	Yes	No
Parsippany, NJ	790	600	4,900	Yes	No	Yes	Yes
Deer Park, NY	1,240	96	8	No	No	Yes	Yes, but Slow
Petaluma, CA	9,900	<2,500	30,000	No	Yes	Yes	Yes
Port Hueneme, CA	4,700	3,300	<5	No	No	Not Tested	Yes
Vandenberg AFB, CA	49,000	3,400	124	No	No	Yes	Yes
Elizabeth City, NC	154	46	1,280	Yes	No	Yes	Not Tested
Nederland,TX	1,500,000	68,000	65	Yes	No	Not Tested	Not Tested

only three of the six sediments tested where there was evidence that MTBE degraded at field scale. In the microcosms where MTBE degraded, there was a near stoichiometric production in TBA (Wilson et al., 2005a).

Novak and his co-workers surveyed anaerobic biodegradation of TBA in sediments from uncontaminated subsurface material from Blacksburg, Virginia; Newport News, Virginia; and Williamsport, Pennsylvania (Novak et al., 1985; Hickman et al., 1989; Hickman and Novak, 1989; Yeh and Novak, 1994). In these studies, TBA degraded readily under nitrate-reducing and sulfate-reducing conditions. Finneran and Lovley (2001) reported degradation of TBA under iron-reducing conditions in sediment from a gasoline spill site in South Carolina and in sediment from the Potomac River. Finneran and Lovely (2003) demonstrated TBA biodegradation under methanogenic conditions in sediment from a refinery spill site in Oklahoma. Day and Gulliver (2003) used long-term monitoring data and analysis of stable carbon isotopes to document natural anaerobic biodegradation of TBA in ground water at a refinery site in Texas.

In contrast to this precedent for TBA biodegradation in the literature, TBA did not degrade in aquifer sediment under anaerobic conditions in studies reported by Suflita and Mormile (1993) and Mormile et al. (1994), and TBA degraded in only one of eight sediments from gasoline spill sites (Table 4.3). At this writing, anaerobic biodegradation of TBA at gasoline spill sites has not been well documented.

The degradation of ethanol was rapid in sediments from the gasoline spill sites at Petaluma, California; Parsippany, New Jersey; and Vandenberg AFB, California; the first order rate of attenuation was greater than 0.1 per day. In sediments from the sites at Deer Park, New York, and Port Hueneme, California, the rate was slower, but still environmentally significant. The rates were greater than 2 per year.

Ethanol did not degrade in sediment from Boca Raton, Florida. The first order rate of degradation was less than 0.05 per year. Acidic conditions can inhibit anaerobic biodegradation of ethanol; however, low pH does not explain the absence of ethanol degradation in the sediment from Boca Raton, Florida. The pH of the sediment from Boca Raton was 6.5. The sediment used to prepare the microcosms had high concentrations of BTEX compounds and trimethylbenzenes. The demand for fixed nitrogen or for sulfate to metabolize the aromatic hydrocarbons may have depleted the supply of the nutrients. The pore water in the microcosms contained approximately 1.0 mg/L of ammonia N, and less 0.3 mg/L of sulfate.

4.5   Acclimation to Anaerobic Biodegradation of MTBE

Prior to 1999, the Local Oversight Program of the Environmental Health Division of the Health Care Agency in Orange County (California) used EPA method 8020 or 8021 for routine monitoring of fuel-derived contaminants. Concentrations of TBA were not reported. In 2000 and 2001, they transitioned their monitoring to EPA method 8260 or 8260B (purge and trap with gas chromatography with a mass spectrometer detector), and concentrations of TBA were routinely reported. The concentrations of TBA were higher than they had expected (Figure 4.2). On average, the concentrations of TBA were higher than the concentrations of MTBE. A similar relationship for TBA and MTBE was reported for neighboring Los Angeles County (Shih et al., 2004).

The Local Oversight Program selected thirteen gasoline spill sites for detailed evaluation. All of the sites had high concentrations of TBA. Some had low concentrations of MTBE, and some had higher concentrations of MTBE (Figure 4.2). In 35 of 59 wells at the sites selected for detailed study, the concentrations of TBA and MTBE followed a pattern illustrated in Figure 4.3. Initially, each monitoring well produced ground water with higher concentrations of MTBE and lower concentrations of TBA. After a period of time, the well started to produce ground water with high concentrations of TBA and much lower concentrations of MTBE. The transition from water that was dominated by MTBE to water that was dominated by TBA was usually rapid, taking only a few months to a year. The decline in concentrations of MTBE and increase in concentrations of TBA occurred in water that was anaerobic. This sharp transition was probably a result of the acclimation of anaerobic microorganisms in the aquifer to degrade MTBE.

The field data in Figure 4.3 are typical of field data from the 35 wells that exhibited the transition from MTBE to TBA. Sometimes more TBA is produced than would be expected from the biodegradation of concentrations of MTBE that were present at an earlier time; sometimes less. This most likely reflects changes in the concentration of MTBE in the ground water before biodegradation, caused by changes in ground water elevation, or changes in flow direction, or some other influence.

This behavior of MTBE at field scale is mirrored in the behavior of MTBE in anaerobic microcosms. Figure 4.4 presents data from a microcosm study conducted with sediment from Parsippany, New Jersey (Table 4.3; Kolhatkar et al., 2002; Wilson et al., 2005a). After a lag, the transition from MTBE to TBA was rapid. The first order rate of attenuation of concentrations of MTBE (including the lag) in the microcosms was 11.7 ± 2.4 per year at 90% confidence. In the microcosm data presented in Figure 4.4, the MTBE that was removed was replaced with an equivalent concentration of TBA, confirming that MTBE was being transformed to TBA.

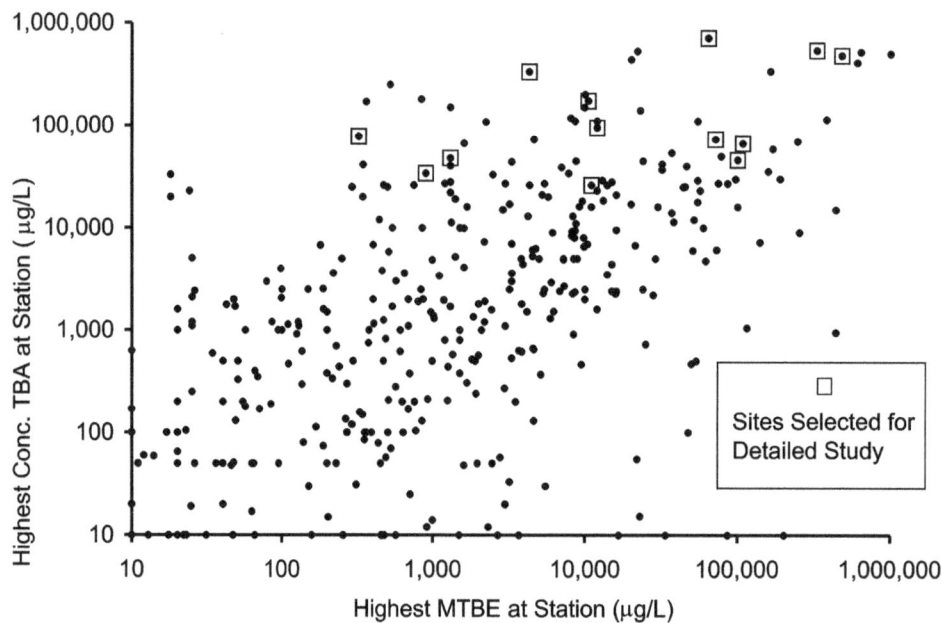

Figure 4.2   *Distribution of TBA and MTBE in the most contaminated wells at gasoline spill sites in Orange County, California, in 2002.*

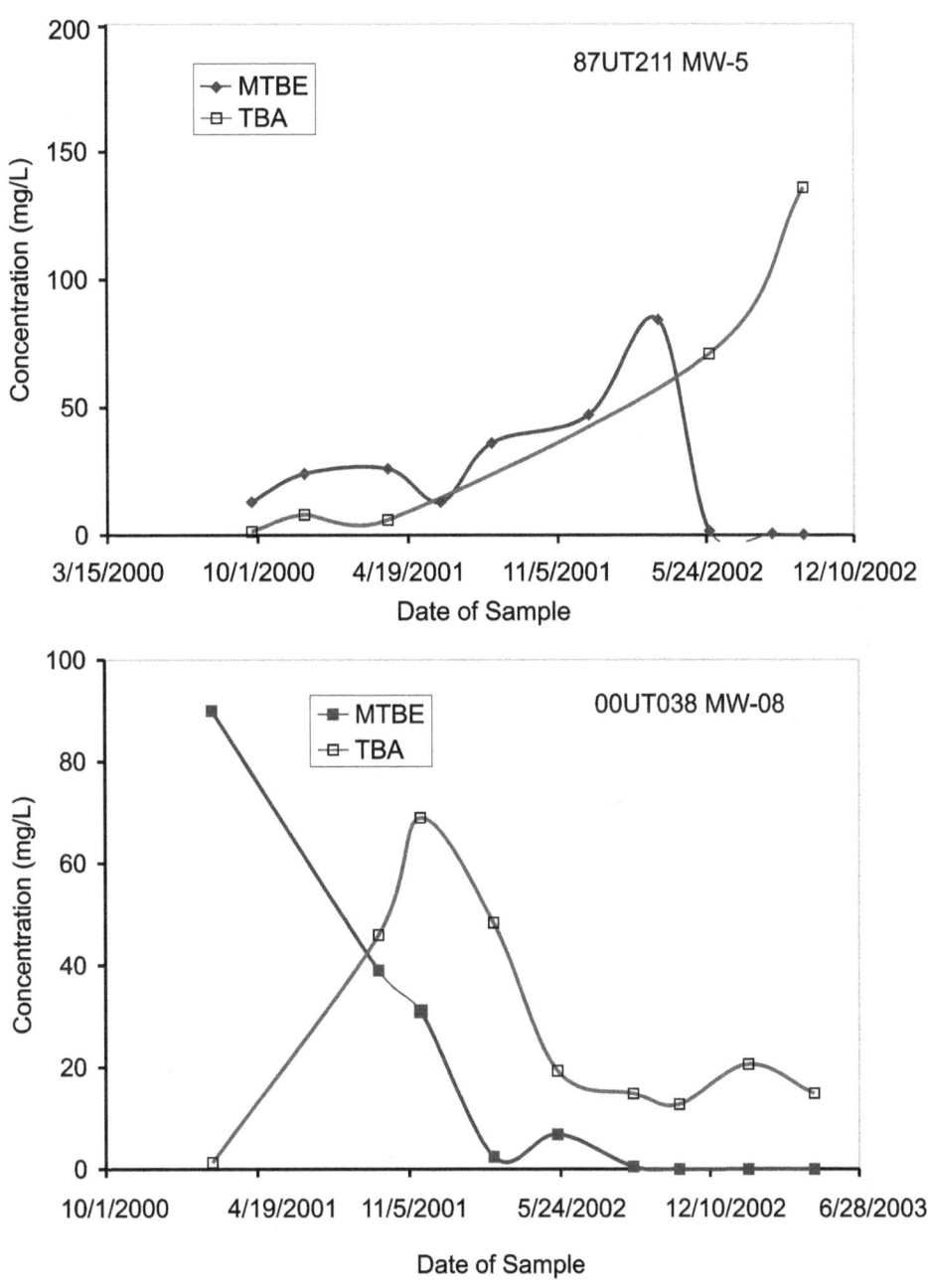

Figure 4.3 *Transition from MTBE to TBA in monitoring wells at gasoline spill sites in Orange County, California.*

The extent of biodegradation of MTBE in the microcosms and enrichment cultures from the microcosms could be predicted from the fractionation of the stable carbon isotopes of MTBE remaining in the microcosm after biodegradation. (Kolhatkar et al., 2002; Kuder et al., 2004). The process is discussed in detail in Section 5. The relationship between the fractionation of the stable carbon isotopes in MTBE and the extent of biodegradation of MTBE in microcosms was used to predict the extent of biodegradation of MTBE in the ground water in Orange County, California (Kuder et al. 2004; Wilson et al., 2005b). The results are presented in Table 4.4. At 12 of the 13 sites, most of the MTBE in the most contaminated well had been biologically degraded. At seven of the thirteen sites, more than 90% of the MTBE had degraded.

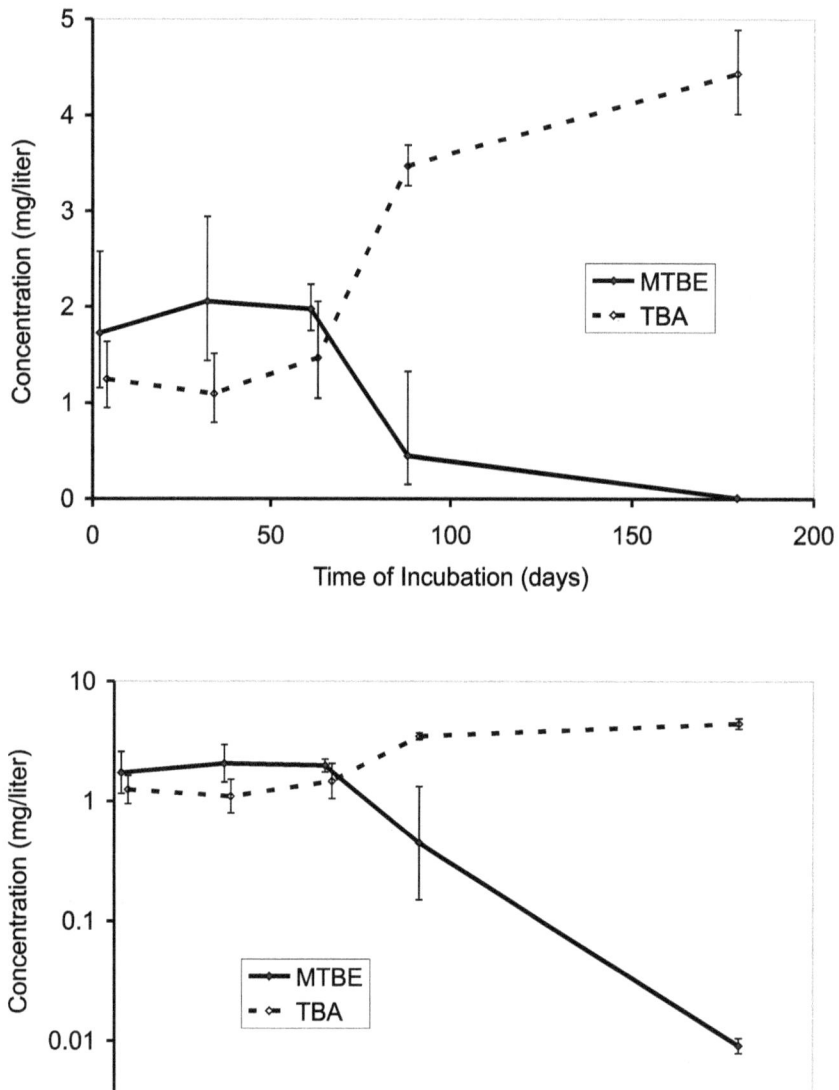

Figure 4.4 -*Anaerobic biodegradation of MTBE and production of TBA in microcosms constructed with sediment from a gasoline spill site. The error bars are the 95% confidence intervals on the geometric mean concentration. The same data are plotted on an arithmetic scale in the upper panel and a logarithmic scale in the lower panel.*

Table 4.4 Biodegradation of MTBE in the Most Contaminated Well at 13 Gasoline Spill Sites in Orange County, California, as Predicted by Fractionation of ^{13}C in MTBE. δ^{13}C is the Unit Used to Measure the Isotopic Fractionation (See Section 5 for a Detailed Explanation.)

Location	MTBE (µg/L)	TBA (µg/L)	δ^{13}C MTBE ($^o/_{oo}$)	Fraction MTBE remaining
99UT015	280,000	300,000	-30.32	Biodegradation not expected
99UT032	2,500	24,000	-18.51	0.478
96UT028	2,650	190,000	-13.29	0.312
87UT211	268	136,000	-12.37	0.289
91UT086	890	81,000	-0.71	0.111
00UT038	5.49	51,600	5.29	0.068
86UT175	100	80,000	6.04	0.064
89UT007	820	29,000	6.84	0.060
86UT062	20	13,000	15.95	0.028
88UT138	16.9	180,000	24.03	0.015
88UT198	100	41,000	27.06	0.011
85UT114	100	110,000	56.78	0.001

4.6 Zero Order Biodegradation of MTBE at High Concentrations -

All the equations used in this report assume that the anaerobic biodegradation of MTBE is a first order process. If the removal is a first order process, the data will plot along a straight line when time or distance is plotted on an arithmetic scale and concentrations of MTBE are plotted on a logarithmic scale. The lower panel in Figure 4.4 plots the data with concentration of MTBE on a logarithmic scale. The concentrations of MTBE fall along a straight line from day 61 of the incubation to day 179 of the incubation, indicating that the anaerobic biodegradation of MTBE in the microcosms was a first order process. The initial concentration of MTBE in the microcosms constructed with sediment from Parsippany, New Jersey, was near 2 mg/L. The anaerobic degradation of MTBE in microcosms constructed with material from a JP-4 spill in Elizabeth City, North Carolina, was also first order (Wilson et al., 2000). The initial concentration of MTBE in the sediment from Elizabeth City, North Carolina was near 3 mg/L. At these concentrations, and at lower concentrations, the anaerobic biodegradation of MTBE can be expected to be a first order process. First order rate constants have a unit of reciprocal time, such as, per day or per year.

At higher concentrations, the enzymes that metabolize MTBE may become saturated. As a consequence, the bacteria degrade MTBE at some fixed maximum rate, regardless of the concentration of MTBE. Under these conditions, the rate of degradation is described as a zero order process. Degradation follows Equation 4.1, where Co is the initial concentration, C is the final concentration, t is the elapsed time, and K is the zero order rate constant. Typical units for K would be mg/L per day.

$$C = Co - Kt$$

Equation 4.1

The microcosms that produced the data in Figure 4.4 were respiked with MTBE at an initial concentration near 100 mg/L. The biodegradation at this higher concentration is presented in Figure 4.5. If biodegradation is zero order, the data should plot along a straight line when the concentrations are plotted on an arithmetic scale. The upper panel in Figure 4.5 indicates that biodegradation proceeded without a lag, and that the degradation was a reasonably good fit to a zero order process.

The lower panel of Figure 4.5 plots the same data on a logarithmic scale. The logarithmic plot suggests that the biodegradation of MTBE behaved like a zero order process for the first four sampling dates, and like a first order process for the remainder of the incubation. In the first four sampling dates, the zero order rate of biodegradation was 0.20 mg/L per day. In the remainder of the incubation, the first order rate of biodegradation was 10 ± 3.4 per year, which is not statistically different from the first order rate achieved by the organisms in the sediment in the experiments described in Figure 4.4.

The transition from zero order to first order biodegradation occurred at concentrations of MTBE between 65 and 40 mg/L. Using concentrations of MTBE above 40 mg/L in Equations 2.2 and 2.3 will likely produce errors, but the errors are conservative. The MTBE will degrade more rapidly than would be predicted by the equations.

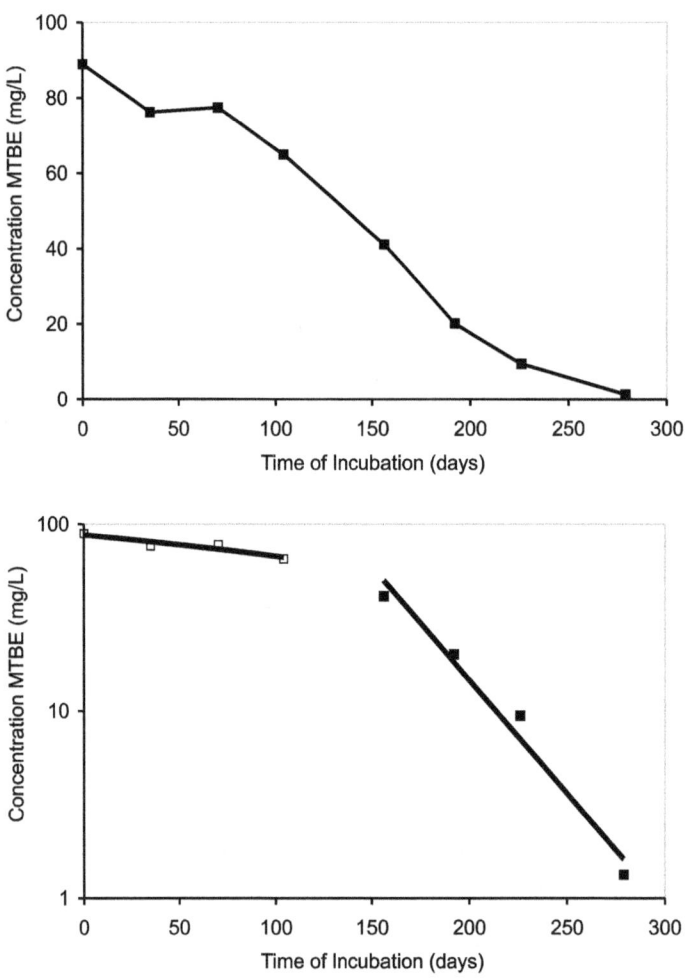

Figure 4.5 -*Anaerobic biodegradation of MTBE at high concentration in an enrichment culture constructed with sediment from the microcosms used to produce the data in Figure 4.4. The same data are plotted on an arithmetic scale in the upper panel and a logarithmic scale in the lower panel.*

Section 5 -

Monitoring MTBE Biodegradation with Stable Isotope Ratios -

A new technique has been developed to evaluate the extent of MTBE biodegradation at field scale. The technique is based on the fractionation of the stable carbon isotopes in the remaining MTBE during the course of degradation. The fractionation of the stable carbon isotopes of MTBE can provide an unequivocal indication of MTBE biodegradation. At this writing, these analyses are only commercially available from a few university laboratories. Their costs are on the order of $250 per analysis. In the future, reports provided to regulators to evaluate the risk from MTBE plumes are likely to contain data on the ratio of stable carbon isotopes in the residual MTBE in the ground water. This is particularly true when the possibility of natural biodegradation of MTBE is crucial to the risk evaluation of a gasoline spill.

These analyses are not necessary at gasoline spill sites where the possibility of MTBE biodegradation does not change the site conceptual model or the strategy for risk management at the site. It is not necessary to evaluate MTBE fractionation in every well at a site. The analyses should be reserved for water from wells that are critical to the risk analysis.

This section provides a visual illustration of the process of isotope fractionation during biodegradation. It also provides numerical examples of the fractionation that can be expected during anaerobic biodegradation. This section explains the units used to express the ratio of stable carbon isotopes and presents simple formulas and graphs to predict the extent of biodegradation from the measured stable carbon isotope ratio.

5.1 Monitoring MTBE Biodegradation with Stable Isotope Ratios

There are two stable isotopes of carbon: carbon twelve (^{12}C) and carbon thirteen (^{13}C). Unlike carbon fourteen (^{14}C), the stable isotopes are not subject to radioactive decay. The most prevalent stable isotope of carbon is ^{12}C. Approximately 1% of the carbon on the Earth is ^{13}C. During MTBE biodegradation, MTBE molecules with ^{12}C at the methoxy group are metabolized more rapidly than MTBE molecules with ^{13}C at the methoxyl group (Hunkeler et al., 2001; Gray et al., 2002; Kolhatkar et al., 2002; Kuder et al., 2004, 2005). This discrimination against the heavier isotope is called the kinetic isotope effect. As biodegradation of MTBE proceeds, the remaining MTBE contains a progressively greater proportion of the ^{13}C isotope. As a consequence, the extent of biodegradation can be determined from the change in the ratio of stable isotopes in the MTBE. Figure 5.1 is a pictorial illustration of this fractionation. The figure greatly exaggerates the relative fractionation compared to ^{12}C and ^{13}C to make it easier to see the effect.

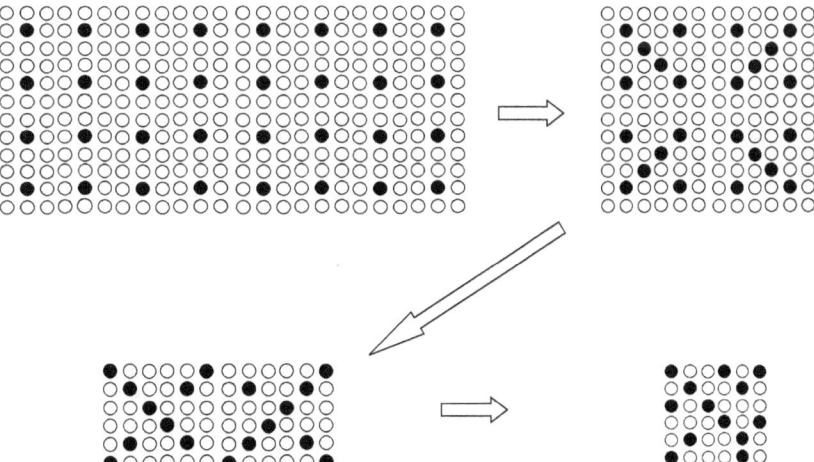

Figure 5.1 *An illustration of the kinetic isotope effect. Shown is the enrichment in black dots, or fractionation of the proportion of black dots to white dots, when the rate of removal of white dots is faster than the removal of black dots.*

The first order rate of biodegradation (k) can be calculated from the initial concentration of MTBE (C_o), the final concentration of MTBE (C), and the time elapsed (t) following Equation 5.1. Equation 5.1 rearranges the terms in Equation 2.2 in Chapter 2.

$$k = -\ln\left(C/C_o\right)/t$$

Equation 5.1

The numbers of black dots and white dots at each time step in Figure 5.1 are presented in Table 5.1. Half of the dots are degraded in one time step. Each time step is one half-life for the dots. From Table 5.1, the number of black dots at time zero was 32, and the number after time step three was 13. If 13 is substituted for C, 32 is substituted for C_0, and 3 is substituted for t in Equation 5.1, the first order rate of attenuation of black dots is 0.300 per time step. Similarly, the rate of attenuation of white dots is 0.782 per time step, and the rate of attenuation of all the dots is 0.693 per time step. The rate of removal of white dots was 2.6 times faster than the rate of removal of black dots, and the black dots became progressively more abundant relative to the white dots with each time step. This visual analogy will be applied to the fractionation of stable carbon isotopes in MTBE during the course of biodegradation.

Table 5.1   Fractionation of Black Dots and White Dots in the Visual Example in Figure 5.1

Time Step (half-lives)	Number of All Dots Remaining	White Dots Remaining	Black Dots Remaining	Ratio Black Dots to White Dots
zero	288	240	32	0.13
one	144	120	24	0.20
two	72	54	18	0.33
three	36	23	13	0.57

Recent advances in analytical chemistry make it possible to determine the ratio of stable isotopes in MTBE dissolved in a water sample at concentrations that are near regulatory standards (Hunkeler et al., 2001; Kolhatkar et al., 2002). The MTBE is separated from water by purge and trap (Kolhatkar et al., 2002) or by solid phase microextraction (Hunkeler et al., 2001), and then further separated by gas chromatography, and finally the ratio determined with an isotope ratio mass spectrometer. The effective minimum concentration of MTBE for analysis of the stable carbon isotope ratio is near 10 μg/L.

The isotope ratio mass spectrometer does not measure the ratio of the stable carbon isotopes directly to each other. Rather, it measures the deviation of the ratio in the sample from the ratio in a standard substance that is used to calibrate the instrument. The substance used as the international standard for stable carbon isotopes has a ratio of ^{13}C to ^{12}C of 0.0112372.

The conventional notation for the ratio of ^{13}C to ^{12}C in a sample ($\delta^{13}C$) reports the ratio in terms of its deviation from the ratio in the standard.

$$\delta^{13}C = \left[\frac{\left(^{13}C/^{12}C\right)_{sample} - \left(^{13}C/^{12}C\right)_{standard}}{\left(^{13}C/^{12}C\right)_{standard}}\right] \times 1000$$

Equation (5.2)

The units for $\delta^{13}C$ are parts per thousand, often represented as ‰ , or per mil, or per mill. Table 5.2 presents calculations that illustrate the changes in $\delta^{13}C$ when carbon in MTBE is fractionated during biological degradation. In the example calculations, the abundance of ^{13}C starts out at 1.092% of the total carbon. This abundance is in the range typically encountered in MTBE in gasoline. The $\delta^{13}C$ calculated following Equation 5.2 is -28.2 ‰. The first order rate of biodegradation of MTBE containing only ^{12}C was 0.6927 per time step, while the rate of biodegradation of MTBE containing one atom of ^{13}C was 0.6852 per time step. The rate of degradation of MTBE containing only ^{12}C is approximately 1.09% faster than the rate of degradation of MTBE containing one atom of ^{13}C. This ratio of the rates of biodegradation is typical for anaerobic biodegradation of MTBE.

After ten time steps, the concentration of MTBE is reduced nearly one thousand fold, and the ratio of ^{13}C to ^{12}C increased from 0.01092 to 0.01182. The value of $\delta^{13}C$ shifts from –28.4 ‰ to +52.2 ‰. This is the range of values of $\delta^{13}C$ in MTBE typically seen at field sites to date. The value of $\delta^{13}C$ can usually be determined with a reproducibility

Table 5.2 Typical Changes in the Ratio of ^{13}C to ^{12}C in MTBE During Biodegradation of MTBE under Anaerobic Conditions

Step	Fraction Remaining MTBE	Fraction Remaining ^{12}C	Fraction Remaining ^{13}C	$^{13}C/^{12}C$	$\delta^{13}C$ (‰)
0	1.000	0.989	0.0108	0.01092	-28.2
1	0.500	0.495	0.00544	0.01101	-20.2
2	0.250	0.247	0.00274	0.01109	-13.1
3	0.125	0.124	0.00138	0.01118	-5.1
4	0.063	0.0618	0.000697	0.01127	2.9
5	0.031	0.0309	0.000351	0.01136	10.9
6	0.016	0.0155	0.000177	0.01145	18.9
7	0.008	0.00773	8.9E-05	0.01154	26.9
8	0.004	0.00386	4.4E-05	0.01164	35.8
9	0.002	0.00193	2.2E-05	0.01173	43.9
10	0.001	0.00097	1.1E-05	0.01182	51.9

of better than ± 0.5‰. The value of $\delta^{13}C$ before biodegradation was negative because MTBE in gasoline has relatively less ^{13}C than does the international standard used to calibrate the mass spectrometer. As biodegradation proceeded, the value of $\delta^{13}C$ became more positive.

Figure 5.2 plots the value of $\delta^{13}C$ in Table 5.2 against the fraction of MTBE remaining after each step. Notice that the extent of biodegradation increases and the fraction of MTBE remaining decreases toward the right-hand side of the x-axis. Table 5.2 contains data that are typical for anaerobic biodegradation of MTBE; Figure 5.2 also plots data typical of aerobic biodegradation. In the calculations for aerobic biodegradation, the first order rate of degradation of MTBE composed entirely of ^{12}C was 0.6927 per time step (the same as for calculations for anaerobic biodegradation). The rate of aerobic biodegradation of MTBE containing one atom of ^{13}C was 0.6915 per time step. The rate of aerobic degradation of MTBE composed entirely of ^{12}C is approximately 0.17% faster than the rate of degradation of MTBE containing one atom of ^{13}C. This ratio of the rates of biodegradation is typical for aerobic biodegradation of MTBE.

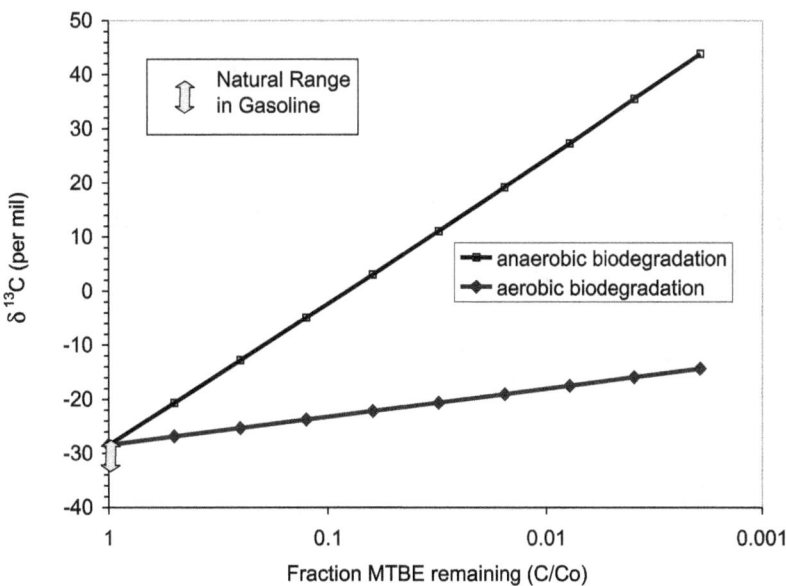

Figure 5.2 *Typical changes in the value of $\delta^{13}C$ as MTBE is degraded under aerobic and anaerobic conditions.*

The rate of anaerobic biodegradation of MTBE containing one atom of ^{13}C in each time step is slower than the rate under aerobic conditions. There will be a greater tendency for ^{13}C MTBE to accumulate under anaerobic conditions, and MTBE will fractionate more rapidly under anaerobic conditions.

These assumptions are consistent with what is known of the physiology of MTBE biodegradation. The initial reaction under anaerobic conditions involves hydrolysis of the ether bond (Kuder et al., 2005, Somsamak et al., 2005). Aerobic metabolism of MTBE is initiated by mono-oxygenase enzymes which extract a proton from the methoxyl group. The difference between the strength of the ^{13}C-O bond and ^{12}C-O bond is more pronounced than the difference between the ^{13}C-H bond and the ^{12}C-H bond (Huskey, 1991), resulting in greater fractionation of carbon during anaerobic metabolism of MTBE.

5.2 Predicting Biodegradation from $\delta^{13}C$ in MTBE in Gasoline

Smallwood et al., (2001) reported that the normal range of $\delta^{13}C$ for MTBE in gasoline is from –28.3‰ to –31.6‰; more recent surveys indicate that the normal range extends between –27.5 ‰ and –33 ‰ (O'Sullivan et al., 2003). This is the range of $\delta^{13}C$ that would be expected for MTBE in ground water in the absence of biodegradation. The variation in $\delta^{13}C$ during anaerobic biodegradation is much larger than the variation in $\delta^{13}C$ in MTBE from one sample of gasoline to another (Figure 5.2).

Notice the straight-line relationship when values of $\delta^{13}C$ on an arithmetic scale are plotted against the fraction of MTBE remaining on a logarithmic scale in Figures 5.2 and 5.3. The simplified version of the Rayleigh equation, originally developed by Mariotti et al. (1981), is commonly used in the literature to relate the extent of biodegradation of MTBE (and other organic compounds) to the $\delta^{13}C$ of the material remaining after biodegradation.

$$\delta^{13}C_{MTBE\,in\,ground\,water} = \delta^{13}C_{MTBE\,in\,gasoline} + \varepsilon \bullet \ln F \qquad \text{Equation 5.3}$$

In Equation 5.3, ε is the isotopic enrichment factor and is an expression of the extent of isotopic fractionation during biodegradation, and F is the fraction of MTBE remaining after biodegradation. The value of F is simply C/Co in Figure 5.2. The value of ε is usually calculated as the slope of a linear regression of $\delta^{13}C$ on the natural logarithm of F (Figure 5.3). Notice that the data in Figure 5.3 are plotted in an unconventional fashion. The natural logarithm of F decreases toward the right-hand side of the x-axis.

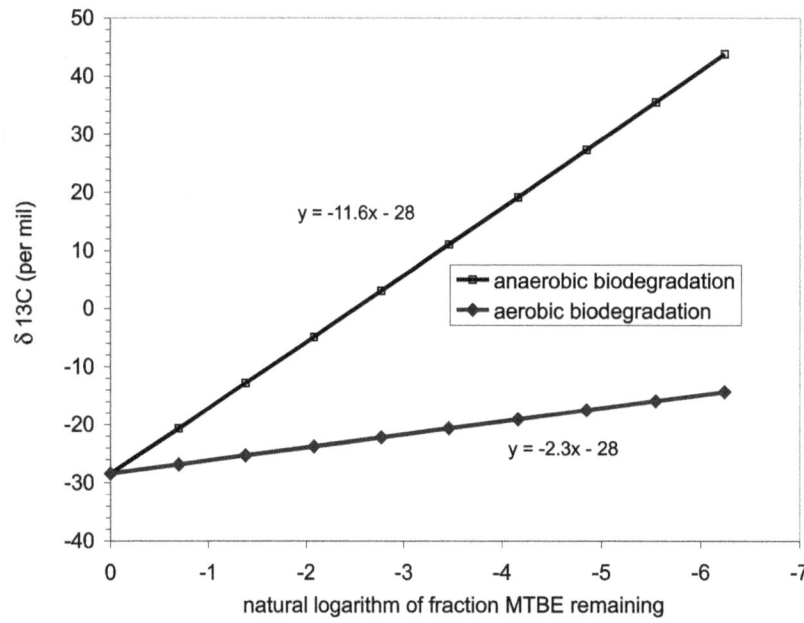

Figure 5.3 *The data from Figure 5.2 have been plotted in units commonly used in the literature on stable isotopes.*

34

Hunkeler et al., (2001) determined the enrichment factor for four aerobic cultures that degraded MTBE as the sole carbon source and one culture that co-metabolized MTBE when grown on 3-methylpentane. All the cultures were enriched from Borden aquifer sediments from Ontario, Canada. The enrichment factors varied from −1.52 ± 0.06 to −1.97 ± 0.05. Gray et al., (2002) determined the enrichment factors during aerobic growth of pure culture PM-1 and a mixed culture from Vandenberg AFB, California. The enrichment factors varied from −1.4 ± 0.1 to −2.4 ± 0.3.

Under anaerobic conditions, the enrichment factors are more negative. Kolhatkar et al., (2002) extracted an enrichment factor of −8.1 ± 0.85 for anaerobic biodegradation along a flow path at a gasoline spill in New Jersey and −9.1 ± 5.0 for degradation in anaerobic microcosms constructed with core material from the site. More recent work indicates that the isotopic enrichment factor in a mixed culture of the organisms from New Jersey is −12.5 ± 1.4 (Kuder et al., 2005). Somsamak et al., (2005) determined the enrichment factor for MTBE biodegradation by an enrichment culture isolated from sediment from the Arthur Kill inlet between Staten Island, New York, and New Jersey. In a methanogenic culture, the value of ε was -15.6 ± 4.1. These experimental values are in good agreement with a theoretical value of ε (−12.2) that would be expected from the cleavage of a C-O bond in a molecule with five carbon atoms (Kuder et al., 2005 following Huskey, 1991).

The values of ε in the hypothetical data in table 5.2 and Figure 5.3 would be −11.6 for anaerobic biodegradation and −2.3 for aerobic biodegradation. The values of ε in the examples were based on typical values derived from laboratory experiments and field studies.

5.3   Sources of Uncertainty in Estimates of Biodegradation

The relationship between the $\delta^{13}C$ of MTBE in a ground water sample and the true extent of biodegradation is often complex. It is influenced by the starting $\delta^{13}C$ of the MTBE in the gasoline spill, and this value is rarely known with certainty. There may be different releases of gasoline with different values of $\delta^{13}C$ contributing to the same spill. The MTBE in a sample of gasoline floating in a monitoring well or in gasoline extracted from a core sample may already be biologically weathered, and the $\delta^{13}C$ may not be representative of the $\delta^{13}C$ of the original release. Because a trustworthy value of $\delta^{13}C$ for the MTBE in the original gasoline in the spill is rarely available, most evaluations compare the $\delta^{13}C$ in the MTBE in the ground water to the published range of $\delta^{13}C$ in samples of gasoline that have not been released to the environment.

The relationship between the value of $\delta^{13}C$ for MTBE and the extent of biodegradation is also influenced by the relative contribution of aerobic and anaerobic biodegradation to the attenuation of MTBE. Because the value of ε is less negative for aerobic biodegradation, there is less fractionation for a given amount of biodegradation. When $\delta^{13}C$ is used to predict the fraction of MTBE remaining, the predicted extent of biodegradation is much greater for aerobic conditions compared to anaerobic conditions. As an example, if the MTBE in the original gasoline spill had a value of $\delta^{13}C$ of −27.5 ‰, and the MTBE in the ground water was −10.0 ‰, the fraction remaining calculated using Equation 5.3 under aerobic conditions would be 0.0006, and the fraction remaining under anaerobic conditions would be 0.20.

The value of $\delta^{13}C$ is also influenced by site specific interactions between the ground water and the source of contamination. If the ground water is in contact with residual gasoline, unfractionated MTBE can dissolve from the gasoline into ground water while biodegradation is in progress. Imagine ground water flowing through and under a region containing residual gasoline. At the leading edge, MTBE dissolves into the ground water. As this MTBE moves with ground water, it is biodegraded and fractionated. As the ground water flows through and underneath the area with residual gasoline, additional MTBE from the center and down gradient edge of area with residual gasoline will dissolve into the ground water. This MTBE will not be fractionated.

The fractionated MTBE remaining in ground water after biodegradation will be diluted by additional MTBE that has not been fractionated. This dilution will shift the value of $\delta^{13}C$ to an extent that is directly proportional to the relative concentrations of fractionated and unfractionated MTBE in the water sample. However, the shift in the value of $\delta^{13}C$ will affect the estimate of biodegradation to an extent that is related to the natural logarithm of the fraction remaining after biodegradation. The overall effect of dilution is to produce a value of $\delta^{13}C$ which will underestimate the true extent of MTBE biodegradation.

A related interaction affects the estimate of biodegradation in a plume that is heterogeneous with depth across the well screen of the monitoring well. If MTBE is almost entirely degraded at one depth interval but not at another, then the MTBE produced from the well will be dominated by water from the interval where MTBE did not degrade as extensively. Fractionated and unfractionated MTBE will be mixed in the monitoring well, and the value of $\delta^{13}C$ would underestimate the true extent of biodegradation in the entire plume.

If the TBA that is produced by biodegradation of MTBE is not further degraded, the next effect of these interactions will be higher concentrations of TBA than would be expected from the extent of MTBE biodegradation as predicted from the value of $\delta^{13}C$ in MTBE in the water sample.

5.4 A Conservative Estimate of the Extent of Biodegradation

These sources of uncertainty make it difficult to predict an exact value for the extent of biodegradation of MTBE in ground water. However, if certain assumptions are made, it is possible to calculate an upper boundary on the fraction of MTBE remaining from the $\delta^{13}C$. In other words, for a certain value of $\delta^{13}C$, we can be certain that the fraction of MTBE remaining is this small or smaller.

This conservative boundary on the extent of biodegradation is presented as the lower line in Figure 5.4. The lower line in the figure is calculated assuming a value of ϵ of -12, which is the most negative value that is theoretically possible. The value of $\delta^{13}C$ in the gasoline was assumed to be -27.5 ‰, which is the highest value that has been reported for MTBE in gasoline. Under these assumptions, a value of $\delta^{13}C$ of 0 ‰ would correspond to C/Co of 0.1 or 90% removal.

The upper line in Figure 5.4 is the boundary calculated at a value of $\delta^{13}C$ in the gasoline of -33 ‰, which is the lowest value that has been reported for gasoline. This line is added to show the effect of uncertainty in the value of $\delta^{13}C$ in the gasoline originally spilled on the predicted extent of biodegradation. A value of $\delta^{13}C$ of 0 ‰ would correspond to C/Co of 0.06 or 94% removal. At a value of $\delta^{13}C$ for MTBE in ground water of 0 ‰, the published range of $\delta^{13}C$ in gasoline would produce a two-fold range in the fraction of MTBE remaining that was predicted from Equation 5.3 or Figure 5.4. The effect is of minor importance at values of $\delta^{13}C$ above 0 ‰.

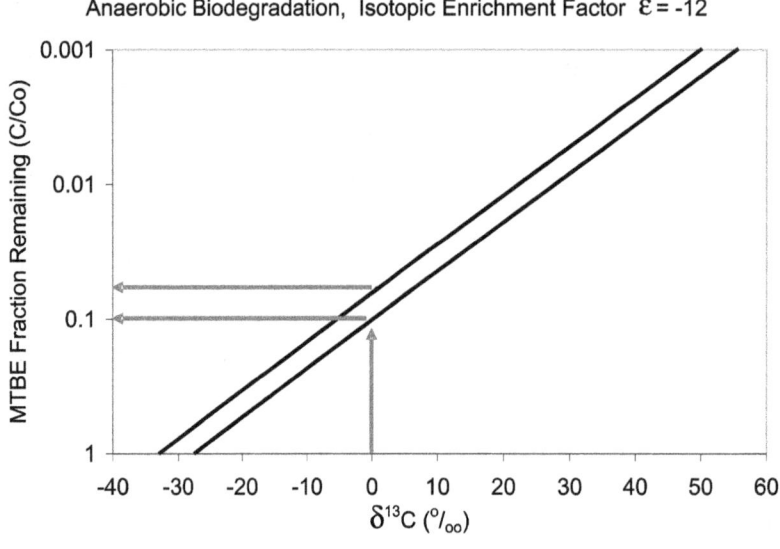

Figure 5.4 *MTBE biodegradation under anaerobic conditions predicted from the $\delta^{13}C$ of MTBE in ground water. The lower line is a conservative prediction of the extent of biodegradation of MTBE.*

Section 6 -

Applications of Stable Carbon Istotope Ratios -

As discussed in Section 5, the stable carbon isotope ratio in MTBE can be used to predict the extent of biodegradation of MTBE in ground water. This section illustrates three practical applications that regulators are likely to see in reports. In the first application, the analysis of stable carbon isotope ratios is used as an alternative to a microcosm study to document that anaerobic biodegradation was occurring at a field site. In the second application, the analyses are used along with data on the hydraulic properties of the aquifer to estimate a first order rate constant for anaerobic biodegradation. In the third application, the analyses are used to show that TBA in ground water came from the biodegradation of MTBE, and not from TBA that was originally present in gasoline.

6.1 Applications of Stable Isotope Ratios to Interpret Plume Behavior

The U.S. EPA recognizes a three-tiered approach to evaluate site specific data in support of monitored natural attenuation (U.S. EPA, 1999), specifically: *(1) historical groundwater and/or soil chemistry data that demonstrate a clear and meaningful trend of decreasing contaminant mass and/or concentration over time at appropriate monitoring or sampling points, (2) hydrogeologic and geochemical data that can be used to demonstrate indirectly the type(s) of natural attenuation processes active at the site, and the rate at which such processes will reduce contaminant concentrations to required levels, 3) data from field or microcosm studies (conducted in or with actual contaminated site media) which directly demonstrate the occurrence of a particular natural attenuation process at the site and its ability to degrade the contaminants of concern. Unless EPA or the overseeing regulatory authority determines that historical data (Number 1 above) are of sufficient quality and duration to support a decision to use MNA, data characterizing the nature and rates of natural attenuation processes at the site (Number 2 above) should be provided. Where the latter are also inadequate or inconclusive, data from microcosm studies (Number 3 above) may also be necessary.*

As a practical matter, it is difficult to provide the second line of reasoning for MTBE in ground water. For any contaminant, it is challenging to obtain the convincing hydrogeologic and biogeochemical data that demonstrate the type of processes and the rate at which the processes operate. The second line of reasoning has been provided in a few instances by correcting the apparent attenuation of the contaminant along a flow path by the attenuation of a tracer. Wiedemeier et al., (1996) were able to use trimethylbenzenes in a plume of ground water contaminated by JP-4 jet fuel as a conservative tracer to correct the apparent attenuation of BTEX compounds along the flow path in the plume. The trimethylbenzenes and BTEX compounds occurred together in the fuel spill in the same relative proportions. Any reduction in concentrations of BTEX compounds in excess of the dilution of the tracer was assumed to be biological degradation. Varadhan et al., (1998) were able to use chloride in a plume of landfill leachate to correct the apparent attenuation of benzene, 1,1-dichloroethane, and 1,2-dichloroethane. Chloride was abundant in the leachate and not in the ambient ground water, and the concentration of chloride was orders of magnitude higher than the concentration of organic chlorine in 1,1-dichloroethane, and of 1,2-dichloroethane.

Unfortunately, at gasoline spill sites, a good tracer for MTBE is usually not available. The biogeochemical parameters, such as depletion of oxygen or production of methane, are associated with gasoline components in general and not MTBE in particular. Any trimethylbenzenes in ground water may have been part of an earlier spill of gasoline that did not contain MTBE. Chloride is not an important component of unleaded gasoline.

Although microcosm studies (the third line of reasoning) are easy to interpret, they are expensive and tend to be time-consuming. Microcosm studies often take several months to over a year to complete, and frequently the results are equivocal. Microcosm studies can only show that the aquifer harbors microorganisms that are capable of degrading the contaminants under the conditions that pertained at the time the aquifer material was sampled. They do not provide direct evidence that the contaminant in the aquifer was actually biologically degraded.

As a consequence, most evaluations of MNA rely heavily on the first line of reasoning, using the long term monitoring data. The rate of decrease in concentration over time in monitoring wells reflects the rate of attenuation of the source of contamination, not the rate of transformation of contaminants along the flow path in ground water (Newell et al.,

2002; Wilson, 2003a; Wilson and Kolhatkar, 2002). Most studies in support of MNA do not provide a solid estimate of the rate of degradation of the contaminant in ground water along the flow path. Without a solid estimate of the rate of degradation in the ground water, a conservative evaluation of the risk to a receptor is restricted to the assumption that the contaminant does not degrade at all.

One conventional approach to evaluate biodegradation of organic contaminants in ground water is to demonstrate an increase in the concentration of transformation products. This approach is problematic for MTBE from gasoline spills because the primary transformation product (TBA) can also be a component of gasoline (compare Landmeyer et al., 1997). Kramer and Douthit (2000) extracted gasoline from six service stations in New Jersey using a fuel-to-water ratio of one to four. TBA was detected in extracts of the gasoline from five of the six stations. The concentration in the water extracts varied from 1,120 to 1,690 mg/l. These are concentrations that would be expected if the MTBE added to the gasoline contained 11% by volume TBA (equivalent to 1.5% by volume in the gasoline). Using TBA to estimate MTBE biodegradation is further complicated by the possibility that TBA may be biologically degraded in ground water under both aerobic and anaerobic conditions.

The Committee on Intrinsic Remediation of the National Academy of Sciences (NRC, 2000) determined that biological transformation was the dominant process responsible for attenuation of MTBE in ground water. Because of these uncertainties, they further determined that the current level of understanding of biological transformation of MTBE is moderate, and as a result, the likelihood of success for using monitored natural attenuation as a remedy for MTBE contamination at a particular site is low (NRC, 2000, page 8).

Recent work shows a strong discrimination during anaerobic biodegradation between molecules of MTBE containing the stable isotope ^{12}C and molecules containing the stable isotope ^{13}C (Kolhatkar et al., 2002; Kuder et al., 2005; Somsamak et al., 2005, see discussion in Section 5). The ^{12}C isotope is preferred, and the molecules containing the ^{13}C isotope accumulate in the residual MTBE. The stable isotopes are said to be fractionated during biodegradation.

6.2   Using Stable Carbon Isotope Ratios to Recognize Natural Biodegradation.

As discussed in Section 5, the extent of degradation of MTBE can be estimated from the change in the ratio of ^{12}C to ^{13}C in MTBE in the plume compared to the ratio in the MTBE in the gasoline that was originally spilled. The stable isotope approach provides direct information on the extent to which MTBE has been biologically degraded in the ground water. Compound-specific stable isotope analysis provides a useful extension to the conventional practice for interpreting the behavior of MTBE plumes.

The approach will be illustrated with data from a plume of MTBE from a gasoline spill site in Dana Point, California (Figure 6.1). The plume of MTBE is contained in a layer of silty fine sand and clean sands that lies beneath a layer of clay and silt. The water table is in the layer of clay and silt. Based on aquifer testing, the average hydraulic conductivity of the layer of silty fine sand and clean sand is 11 meters per day.

The direction of ground water flow at the site was estimated by using linear regression to fit a plane to the elevation of the water in 14 monitoring wells at the site. The regression was fit using the Optimal Well Locator (OWL) application (Srinivasan et al., 2004). Data were available for 14 rounds of sampling between April 1999 and July 2002. For six of the regressions, the value of r^2 was low (0.17 or less), and data from these dates were ignored. For the remaining eight dates, the value of r^2 for the regression ranged from 0.62 to 0.73. For these eight dates, the direction of ground water flow is presented by the "flow rose" in Figure 6.1. The length of each arrow was calculated by multiplying the hydraulic gradient by the average hydraulic conductivity (11 meters per day), then dividing by an estimate of porosity (0.25).

There were two sources of MTBE contamination in the ground water at this site. The major source was associated with leaking under ground storage tanks (Figure 6.1). The tanks and the surrounding fill material were excavated. Residual gasoline in the aquifer acts as a continuing source of MTBE in ground water. The highest concentrations of MTBE are immediately down gradient of the underground storage tanks. A second source is associated with the distribution lines to the eastern dispenser island. In the wells that are side gradient and far down gradient of the underground storage tanks and the dispenser island, the concentrations of MTBE are lower.

To estimate the fraction of MTBE remaining after biodegradation from the δ^{13}C of MTBE in water from the wells, Equation 5.3 was solved for the fraction remaining to produce Equation 6.1.

$$F = C\Big/ Co = e^{\left(\left(\delta^{13}C_{MTBE\ in\ ground\ water} - \delta^{13}C_{MTBE\ in\ gasoline}\right)\big/ \varepsilon\right)} \qquad \text{Equation 6.1}$$

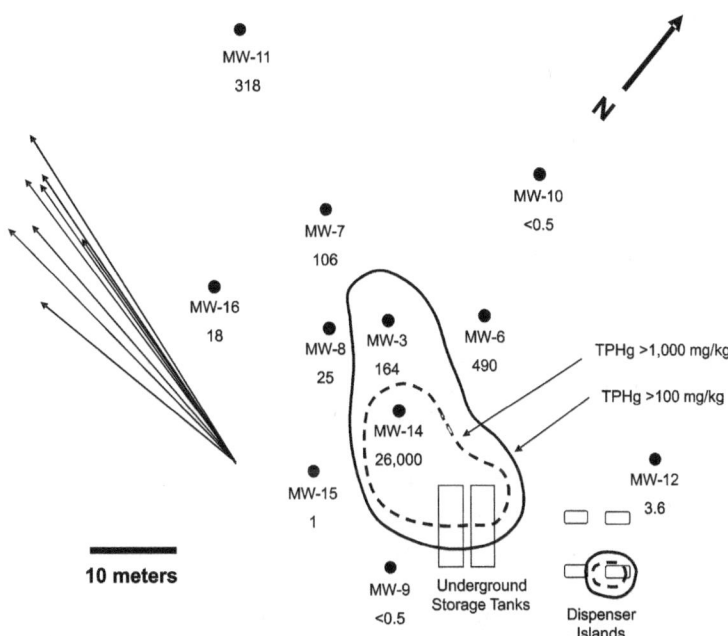

Figure 6.1 *Concentration of MTBE (μg/L) in selected monitoring wells at a gasoline spill site in Dana Point, California, in August 2004. The cluster of arrows is a "flow rose" indicating the direction and distance ground water would move in one year based on the elevation of the water table in monitoring wells on particular sampling dates. TPHg is total petroleum hydrocarbons within the range of molecular weights expected for gasoline.*

Table 6.1 compares the concentrations of MTBE and TBA in the monitoring wells to the fraction of MTBE remaining as predicted from the $\delta^{13}C$ of MTBE using Equation 6.1. A conservative value for ε of -12 was used in the calculation.

Table 6.1 A Comparison Between the Distribution of MTBE and TBA in Ground Water Contaminated by a Fuel Spill in Dana Point, California, and the Extent of MTBE Biodegradation Predicted from the Stable Carbon Isotope Ratio ($\delta^{13}C$) of the Residual MTBE

Well	Date	TBA Measured (μg/L)	MTBE Measured (μg/L)	$\delta^{13}C$ MTBE (‰)	MTBE Fraction Remaining (C/Co)
MW-14	5/20/03	13,000	11,000	-23.88	0.75
	8/18/04	107,000	26,000	-21.58	0.62
MW-3	5/20/03	20,000	870	6.84	0.058
	8/18/04	32,000	164	8.53	0.050
MW-8	5/20/03	10,000	19	18.11	0.023
	8/18/04	32,000	25	37.99	0.0043
MW-6	5/20/03	3,600	47	9.83	0.045
	8/18/04	19,200	490	-1.58	0.116
MW-7	8/18/04	1,220	106	-27.33	0.994
MW-11	5/20/03	<10	1	-31.5*	1.41
	8/18/04	135	318	-28.92	1.14

* *The concentration MTBE was below the limit for the accurate determination of $\delta^{13}C$; the precision of the estimate of $\delta^{13}C$ was ±3 ‰ rather than ± 0.1 ‰.*

Biodegradation makes the value of $\delta^{13}C$ larger. The highest value of $\delta^{13}C$ that has been measured for MTBE in gasoline is -27.4 ‰. This conservative value was used for $\delta^{13}C_{\text{MTBE in gasoline}}$ to calculate the fraction of MTBE remaining of the MTBE originally spilled in the aquifer.

The most contaminated well at the site (MW-14 in Figure 6.1) is located in an area that had 9,000 mg/kg of gasoline range Total Petroleum Hydrocarbons (TPHg). When sampled in May 2003, MTBE in water from MW-14 had a low value of $\delta^{13}C$. The concentrations of MTBE and TBA were essentially equivalent, and there was little evidence of biodegradation (Table 6.1). When sampled again in August 2004, the value of $\delta^{13}C$ was slightly higher, and the concentration of TBA was now four fold higher than the concentration of MTBE. Biodegradation was beginning to influence the distribution of MTBE and TBA in well MW-14.

Wells MW-3 and MW-8 are further down gradient of the source of MTBE that was associated with the underground storage tanks. The sum of the concentrations of MTBE and TBA in wells MW-3 and MW-8 are roughly equivalent to the sum of MTBE and TBA in well MW-14; however, the concentration of MTBE is much lower than the concentration of TBA in wells MW-3 and MW-8, indicating that MTBE may have been degraded to TBA. The $\delta^{13}C$ of MTBE in wells MW-3 and MW-8 is much heavier than MTBE in gasoline (Table 6.1). The calculated fraction of MTBE remaining corresponds to 94% to 99.6% biodegradation of MTBE. The attenuation in concentration of MTBE in wells MW-3 and MW-8 compared to well MW-14 can safely be attributed to biodegradation.

Well MW-6 appears to be side gradient to the source of MTBE associated with the underground storage tanks (Figure 6.1). However, well MW-6 is directly down gradient of the secondary source associated with the dispenser islands. The behavior of MTBE in well MW-6 is very similar to wells MW-3 and MW-8. Concentrations of MTBE are low, and concentrations of TBA are high. The $\delta^{13}C$ of MTBE is high compared to MTBE in gasoline, and the predicted fraction remaining corresponds to 88% to 96% biodegradation of MTBE.

Wells MW-7 and MW-11 are even further down gradient of the source of MTBE. The concentrations of MTBE are low, and it would be tempting to attribute the low concentrations to biodegradation. However, the $\delta^{13}C$ of MTBE in these wells is even lower than the $\delta^{13}C$ in MW-14, the most contaminated well. As discussed in Section 5, the expected range of $\delta^{13}C$ for MTBE in gasoline is –27.5‰ and –33‰ (O'Sullivan et al., 2003). In fact, the $\delta^{13}C$ of MTBE in these wells falls near or within the range of $\delta^{13}C$ expected for gasoline. There is no evidence from the $\delta^{13}C$ of MTBE that biodegradation contributed to attenuation of MTBE in these wells.

6.3   Using Stable Carbon Isotope Ratios to Estimate the Projected Rate of Natural Biodegradation.

Because the $\delta^{13}C$ of MTBE in ground water provides a direct estimate of the fraction of MTBE remaining after biodegradation, it can be used to extract an estimate of the rate of natural biodegradation of MTBE along the flow path. Earlier approaches to extract rate constants from field data used conservative tracers to correct for dilution (Wiedemeier et al., 1996; Varadhan et al., 1998) or made an estimate of the attenuation due to dilution from dispersion (Buscheck and Alcantar, 1995). Because the $\delta^{13}C$ of MTBE in a sample of ground water is not changed by dilution, the ratio C/Co estimated from the $\delta^{13}C$ of MTBE is not changed by dilution, and there is no need to correct for dilution in the estimate of biodegradation. The rate constant can be calculated directly from the fraction of MTBE remaining as estimated from the $\delta^{13}C$ of MTBE, the distance between wells, and an estimate of the interstitial seepage velocity. The projected rates of biodegradation are not equivalent to overall rates of natural attenuation because they do not include the contribution from dilution and dispersion, or sorption.

The projected rate of biodegradation can be expressed directly as a first order rate of removal with distance, or the rate of removal with distance can be multiplied by an estimate of the seepage velocity of ground water to calculate a rate of removal with time of travel (Newell et al., 2002). The projected rate of biodegradation with distance is calculated following Equation 6.2. Biodegradation with time follows Equation 6.3.

$$k_{\text{with distance}} = -\ln(F)/d \qquad \qquad \text{Equation 6.2}$$

$$k_{\text{with time}} = -\ln(F) * v/d \qquad \qquad \text{Equation 6.3}$$

In Equations 6.2 and 6.3, k is the projected rate of natural biodegradation, F is the fraction of MTBE remaining as estimated from $\delta^{13}C_{\text{field}}$ using Equation 6.1, and d is the distance along the flow path between the up gradient well and the down gradient well, and v is the ground water seepage velocity.

The average hydraulic conductivity at the site in Dana Point, California, is 11 meters per day. The average hydraulic gradient over eight rounds of sampling was 0.0023 meter per meter. Assuming the effective porosity is 0.25, the aver-

Table 6.2 - Rates of Natural Biodegradation of MTBE Projected along a Flow Path in Ground Water to Monitoring Wells. Projected Rates Were Calculated from the Estimated Seepage Velocity of Ground Water and the Fraction of MTBE Remaining After Biodegradation. Projected Rates of Biodegradation are not Equivalent to Overall Rates of Natural Attenuation, Because They do not Include Contributions from Dilution and Dispersion, or Sorption -

Well	Date Sampled	Fraction MTBE Remaining (C/C_o)	Distance from MW-14 (meters)	Projected Rate of Biodegradation with Distance (per meter)	Projected Rate of Biodegradation with Time (per year)
MW-3	May, 2003	0.058	9.6	0.30	10.9
MW-3	August, 2004	0.050	9.6	0.31	11.5
MW-8	May, 2003	0.023	11.7	0.32	11.9
MW-8	August, 2004	0.0043	11.7	0.46	17.1
MW-7	August, 2004	0.994	23.0	0.00025	0.0093
MW11	August, 2004	1.0	44.1	0	0
			Distance from Dispenser Island (meters)		
MW-6	May, 2003	0.045	31.1	0.10	3.7
MW-6	August, 2004	0.116	31.1	0.069	2.6

age ground water seepage velocity should be near 37 meters per year. The projected rates of biodegradation of MTBE along flow paths between the most contaminated well (MW-14), and down gradient wells MW-3, MW-7, MW-8, and MW-11 are presented in Table 6.2.

In wells MW-3 and MW-8, the projected first order rate of biodegradation is rapid, on the order of 0.3 per meter of travel, or 10 per year of residence time. In well MW-7, the projected rate of biodegradation was one thousand fold slower, and in well MW-11 biodegradation was not detected at all.

Projected rates along the flow path are most useful to predict the possible extent of plumes, as will be discussed in the following paragraph. The projected rate of biodegradation with time is more convenient to compare the behavior of the plume to other plumes, or to rates published in the literature. The rate of anaerobic biodegradation of MTBE in a microcosm study constructed with material from a gasoline spill in Parsippany, New Jersey, varied from 11 ± 2.3 per year to 12 ± 2.9 per year (Wilson et al., 2005a). The rate of anaerobic MTBE biodegradation in a microcosms study constructed with core material from a JP-4 jet fuel spill in Elizabeth City, North Carolina, was 3.02 ± 0.52 per year and 3.5 ± 0.65 per year (Wilson et al., 2000). These laboratory rates are in reasonable agreement with the rates projected by Equation 6.3 for the flow path to wells MW-3, MW-6 and MW-8 at the Dana Point, California, site (Table 6.2).

If the aquifer carrying the plume is heterogeneous (and most are), it is best to use the highest value for the hydraulic conductivity measured at the site to estimate the seepage velocity used to calculate the projected first order rate with respect to time. It is most likely that the plume is spreading the fastest through the most conductive material. Using the highest value for hydraulic conductivity will provide a conservative estimate of the projected rate constant.

6.4   Using the Projected Rate of Biodegradation to Estimate the Length of Plumes

The distance traveled before the concentration reaches a particular goal (d $_{goal}$) can be calculated by rearranging Equation 6.2 to produce Equation 6.4, where F is the ratio of the goal to the existing concentration in the monitoring well.

$$d_{goal} = -\ln(F)/k_{with\ distance} \qquad\qquad \text{Equation 6.4}$$

If the maximum concentration of MTBE in monitoring wells MW-3 and MW-8 is 1,000 µg/L (compare Table 6.1 for real monitoring data), and the goal for MTBE is the U.S. EPA advisory limit of 20 µg/L, and the projected first order rate of biodegradation with distance is 0.3 per meter; then the plume would be expected to move only 13 meters further before it reaches the goal.

In well MW-7, the projected first order rate of biodegradation is much slower. At a rate of 0.00025 per meter, starting at a concentration of 106 µg/L, the MTBE plume would be expected to move 6,700 meters further down gradient before it reaches the advisory limit. In well MW-11, biodegradation of MTBE could not be established based on the $\delta^{13}C$ for MTBE in the ground water. The only processes that can be reasonably expected to attenuate MTBE further down gradient of wells MW-7 and MW-11 are dilution and dispersion.

This pattern has been seen by the authors in three other MTBE plumes. The biodegradation of the MTBE in the core of the plume was rapid and extensive, but MTBE in the periphery of the plume was not degraded. As a consequence, the extent of the plume was underestimated when a single rate constant for biodegradation was applied to the maximum concentration of MTBE in the source area. On the other hand, the maximum extent of the plume was seriously overestimated if biodegradation was ignored. At this point in the evolution of risk evaluation, a conservative course of action is to recognize that plumes are heterogeneous. An independent estimate of the extent of MTBE contamination further down gradient should be made for each well used in the risk evaluation, based on the concentration of MTBE in each well, and the projected rate of biodegradation in the flow path leading to each well. For many flow paths, the extent of MTBE contamination will be determined by dilution and dispersion, not by anaerobic biodegradation.

6.5 Using $\delta^{13}C$ to Distinguish the Source of TBA in Ground Water

There are two plausible sources of TBA in ground water (compare Landmeyer et al., 1997). Commercial MTBE may have contained as much as 1% to 10% TBA in the past (Kramer and Douthit, 2000). One process for the chemical synthesis of MTBE produces MTBE by reacting isobutylene with methanol. Any water that is present in the methanol feed stock will react with isobutylene to produce TBA, which is carried over into the commercial grade MTBE. In addition, biodegradation of MTBE may produce TBA as a transformation product.

Regulators often need to know the source of TBA in ground water. Tank owners may be reluctant to accept responsibility for a plume that contains high concentrations of TBA if there were low concentrations of TBA in their gasoline. Pump-and-treat can effectively remove TBA from ground water in the source area at a gasoline spill. However, if MTBE continues to partition to ground water from residual gasoline in the aquifer, and the MTBE is degraded to TBA, the concentrations of TBA can rebound. Unfortunately, it is usually impossible to identify the source of TBA present in ground water using conventional chemical analyses.

Figure 6.2 depicts the location of monitoring wells at a gasoline spill site in Delaware that had high concentrations of TBA in the ground water. Well MW-1 is near and slightly up gradient of the underground storage tanks. Well MW-2 is side gradient of the underground storage tanks and down gradient of the dispenser islands. Well MW-3 is down gradient of the underground storage tanks, and MW-4 is further down gradient of MW-3.

Table 6.3 presents the concentrations of contaminants and biogeochemical parameters. A background well (not shown on Figure 6.2) was devoid of contaminants and had low concentrations of methane and moderate concentrations of oxygen and sulfate. The two wells closest to the underground storage tanks (MW-1 and MW-3) had high concentrations of benzene, MTBE, and TBA. The ratio of TBA to MTBE was high. The concentrations of methane were high, and the concentrations of sulfate were low in both wells, and oxygen was depleted in one of the wells. The chemistry of the ground water in these two wells indicates that the water was anaerobic and that natural biodegradation would proceed through the anaerobic pathway.

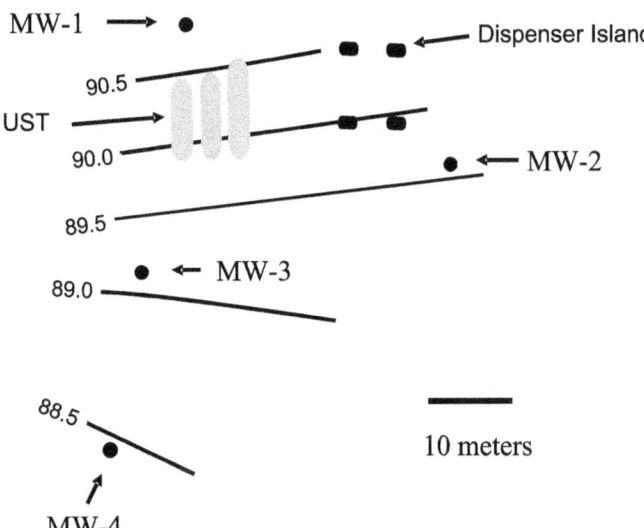

Figure 6.2 *Location of monitoring wells and water table elevations at a gasoline spill site in Newark, Delaware, with high concentrations of TBA in the ground water.*

Table 6.3 Relationship Between the Extent of Contamination and Biogeochemical Parameters at a Site in Newark, Delaware

Location	Benzene	MTBE	TBA	Oxygen	Sulfate	Methane
	(µg/L)			(mg/L)		
Background	<0.5	<1.0	<10	2.7	10.3	0.003
MW-1	1,300	475	245,000	0.5	<0.1	5.4
MW-2	<0.2	45.2	49.4	0.2	4.4	7.1
MW-3	1,440	18,000	306,000	1.6	0.4	2.7
MW-4	<0.5	15.9	406	3.2	24.4	0.06

Water from well MW-2 is also decidedly anaerobic, the concentrations of the contaminants are much lower, and the concentration of TBA is almost equal to the concentration of MTBE. Water from well MW-4 is aerobic, the concentration of contaminants is low, but the concentration of TBA is higher than the concentration of MTBE.

Table 6.4 compares the $\delta^{13}C$ of MTBE in the four wells. The $\delta^{13}C$ of MTBE in well MW-2 is very near the center of the range of values expected for gasoline. There is no evidence of MTBE biodegradation at this location. The value of $\delta^{13}C$ in MTBE in the other three wells is higher than of $\delta^{13}C$ in MTBE in well MW-2.

Table 6.4 Concentrations of TBA Predicted from Biodegradation of MTBE to TBA at a Site in Newark, Delaware

Location	MTBE Measured (µg/L)	$\delta^{13}C$ of MTBE (‰)	MTBE C/Co Figure 5.4	MTBE C/Co Equation 6.1	Prediction of TBA Produced (µg/L)	TBA Measured (µg/L)
MW-1	475	43.05	0.002	0.0021	190,000	245,000
MW-2	45.2	-30.17	1	1.0	0	49.4
MW-3	18,000	7.94	0.05	0.042	350,000	306,000
MW-4	15.9	10.22	0.04	0.035	370	406

The value of $\delta^{13}C$ for MTBE was used to predict the fraction of MTBE remaining in two ways. The estimate of C/Co was estimated graphically from the lower line of Figure 5.4. The fraction remaining (C/Co or F) was also calculated from Equation 6.1.

The value of $\delta^{13}C$ in MTBE measured in MW-2 was assumed to be the value of $\delta^{13}C$ in the gasoline that was originally spilled. Because natural biodegradation of MTBE occurred under anaerobic conditions, the value of ε was assumed to be –12. Results of the calculations are presented in Table 6.4. In the three wells with high values of $\delta^{13}C$, 95% to 99.8% of the MTBE was degraded to TBA.

The amount of TBA that was expected from the biodegradation of MTBE was calculated following Equation 6.5.

$$C_{TBA\,produced} = \left(C_{MTBE}/F\right)\left(1-F\right)\left(74/88\right)$$ Equation 6.5

The measured concentration of MTBE is the concentration after biodegradation. The concentration of MTBE before biodegradation is Co in the expression C/Co. The concentration of MTBE before biodegradation was calculated by dividing the measured concentration by the fraction remaining (C/[C/Co] =Co). The fraction of the original concentration of MTBE before biodegradation that was transformed to TBA is one minus the fraction of MTBE remaining after biodegradation or (1-F). The concentration of MTBE that was transformed to TBA was calculated by multiplying the concentration of MTBE before biodegradation by the fraction of MTBE transformed to TBA. One molecule of MTBE produced one molecule of TBA. The concentration of TBA that was produced was calculated by multiplying the concentration of MTBE that was transformed to TBA by the molecular weight of TBA and dividing by the molecular weight of MTBE ([TBA] = [MTBE] • (74/88)).

Table 6.4 compares the estimated concentration of TBA produced from biodegradation of MTBE to the actual concentration of TBA. The good agreement between the measured concentration of TBA and the expected concentration of TBA indicates that biodegradation of MTBE was responsible for the major portion of the TBA that was present in wells MW-1 and MW-3. Although the concentration of TBA in well MW-4 is lower than the concentrations in MW-1 and MW-3, the TBA that was present in well MW-4 was also produced by biodegradation of MTBE.

The distribution of MTBE at this site is counter intuitive. The MTBE in well MW-2 has not been degraded, but the concentration of MTBE is very low. More than 99% of MTBE in well MW-3 has been degraded, but MW-3 has the highest concentration of MTBE in any well at the site. This disparity can be attributed to heterogeneity in anaerobic biodegradation. The concentrations of MTBE at MW-2 may have been too low to allow acclimation of an MTBE degrading microbial community. As was discussed earlier for the site in Dana Point, California, the core of the MTBE plume participated in anaerobic biodegradation, while flow paths at the dilute margins of the plume did not.

6.6 Caveats and Limitations Concerning the Use of $\delta^{13}C$ of MTBE to Estimate Biodegradation

As discussed in Section 5, a variety of processes operates at field scale to confound the simple relationship in Equation 6.1 between the $\delta^{13}C$ of MTBE in ground water and the extent of biodegradation of MTBE. Some portion of the MTBE may be degraded through an aerobic pathway. Because the value of ε for aerobic metabolism is on the order of -2.5, compared to -12 for anaerobic pathway, a given amount of biodegradation produces a smaller shift in the value of $\delta^{13}C$ of MTBE.

If the MTBE is degraded in ground water that is in proximity to residual gasoline, then fresh MTBE can partition from the gasoline to ground water and dilute the fractionated MTBE with MTBE that has not been fractionated. This effect will be most important near the source areas of plumes, particularly in wells that contained free product at some time in the past.

As illustrated with the data from the site at Dana Point, California, biodegradation in MTBE plumes can be heterogeneous. If one portion of a plume has degraded and a second portion has not, and the two portions are mixed when the water is sampled from a monitoring well, the MTBE in the water from the well will be dominated by MTBE from the portion that did not degrade. The blended value of $\delta^{13}C$ of MTBE in the well water will not accurately reflect the extent of biodegradation of all the MTBE originally present in the ground water.

All of these processes act to underestimate the extent of biodegradation. As a consequence, the fraction remaining calculated by Equation 6.1 is a conservative upper boundary on the fraction remaining. The fraction remaining may be much lower, and the extent of biodegradation of MTBE to TBA may be much larger.

To illustrate this point, Kuder et al., (2004) estimated the fraction of MTBE degraded from the measured concentrations of MTBE and TBA, and then compared the fraction remaining to the $\delta^{13}C$ of MTBE in the ground water. Their results are presented in Figure 6.3. If one molecule of MTBE is degraded to one molecule of TBA, and the TBA is not further degraded, the sum of the molar concentrations of MTBE and TBA after biodegradation should equal the molar concentration of MTBE before biodegradation. The fraction remaining is simply the molar concentration of MTBE after biodegradation divided by the sum of the molar concentrations of MTBE and TBA after biodegradation.

The data presented in Figure 6.3 were collected from 99 wells at 19 sites. The two solid lines are the relationship that would be expected between the fraction of MTBE remaining and the $\delta^{13}C$ of MTBE in ground water if the value of ε is -12, and the value of $\delta^{13}C$ before biodegradation is -33‰ and -27.5‰. Almost all the estimates of the fraction of MTBE remaining based on accumulation of TBA are smaller than the fraction remaining estimated from the $\delta^{13}C$ of MTBE (above the lines in Figure 6.3). The $\delta^{13}C$ underestimated the extent of biodegradation.

The absence of evidence for a process is not evidence for the absence of a process. The approach outlined in Section 5 and illustrated in this section contains a number of conservative assumptions. These include the assumption that all biodegradation of MTBE goes through the anaerobic pathway, and that the $\delta^{13}C$ of MTBE in gasoline that was spilled at any site is as heavy as the heaviest value of $\delta^{13}C$ that has been measured for MTBE in gasoline at any time anywhere in the world. Equation 6.1 may fail to detect natural biodegradation of MTBE when it is really occurring. The situation is directly analogous to a "not detected" in analytical chemistry. A "not detected" does not mean the analyte was not present. If the stable carbon isotope data fail to predict natural biodegradation of MTBE, they should not be further interpreted. In particular, the stable isotope data should not be interpreted to prove that natural biodegradation is not occurring.

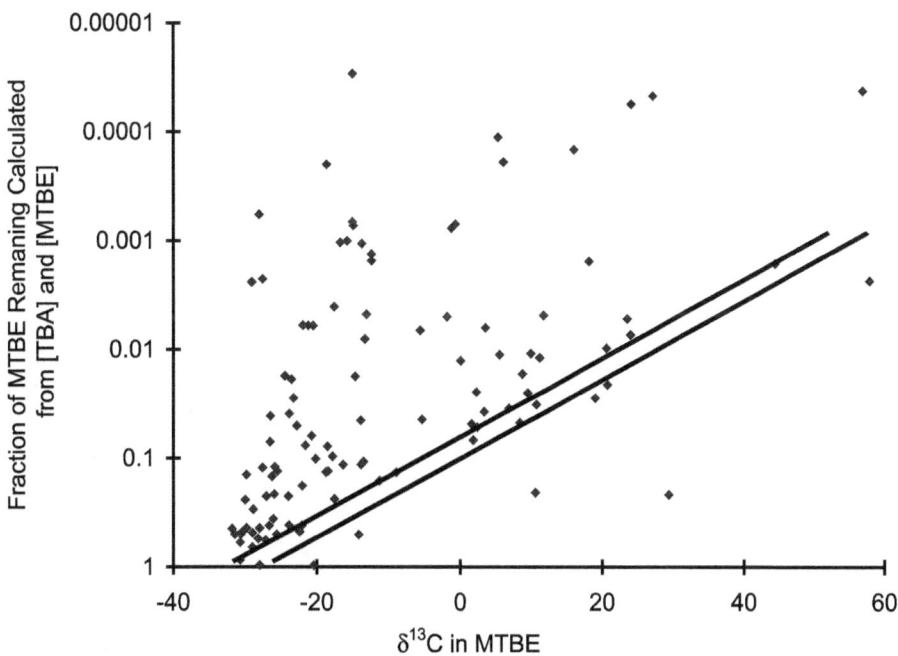

Figure 6.3 *Distribution of $\delta^{13}C$ of MTBE in ground water and the fraction of MTBE remaining from biodegradation as calculated from the concentrations of MTBE and TBA in ground water and the assumption that TBA was produced by biodegradation of MTBE (from Kuder et al., 2005). The solid lines bound the fraction remaining as calculated from the $\delta^{13}C$ of MTBE in ground water.*

Section 7 -

Statistical Evaluation of Rates of Attenuation of Sources -

7.1   Risk Management and U.S. EPA Expectations for MNA

The expectations of the U.S. EPA for natural attenuation in ground water are specified in the OSWER Directive 9200.4-17P (U.S. EPA, 1999). The Directive identifies the first line of evidence for MNA as *historical groundwater and/or soil chemistry data that demonstrate a clear and meaningful trend of decreasing contaminant mass and/or concentration over time at appropriate monitoring or sampling points.* Further, the OSWER Directive 9200.4-17P (U.S. EPA, 1999) notes that *EPA expects that MNA will be an appropriate remediation method only where ... it will be capable of achieving site-specific remediation objectives within a timeframe that is reasonable compared to other alternatives.*

Most states agencies choose to manage gasoline spill sites with a combination of risk management, active clean up, and monitored natural attenuation. The state agencies start with a risk evaluation. If a supply of drinking water is at risk, most state agencies will require active efforts to control the source of MTBE contamination. The risk is reduced through a variety of techniques to remove the gasoline (e.g. free product recovery, excavation of the residual gasoline, and surfactant flushing) and a variety of techniques to treat the gasoline in situ (e.g. air sparging and vacuum extraction, or in situ bioremediation, or electrical heating). Most state agencies focus their effort on source reduction versus remediation of the plume in ground water.

Most state agencies monitor concentrations of MTBE and other fuel components at gasoline spill sites on a fixed schedule. This monitoring has two purposes. It provides documentation that the concentrations in the plume are actually declining over time. It also monitors the plume for a radical change in its behavior that would require a new evaluation of risk. There may be a new release of gasoline at the site. The direction of ground water flow from the spill may have changed due to changes in pumping of ground water from the aquifer or development of land in the recharge zone of the aquifer.

If the trend in contaminant concentrations is down, and the state agency is satisfied that the risk of exposure is properly managed at a site, they may not require an active remedial technology for the site. A clear and meaningful trend can be documented with conventional parametric statistics such as the slope of a regression line or non-parametric statistics such as the Mann-Kendall test.

If the goal is simply to establish that the concentrations are declining over time, either the parametric or non-parametric statistics are useful and appropriate. If the goal is to determine how rapidly the concentrations are declining, or to project how soon the concentration will reach a particular goal, then it is necessary to use parametric statistics.

Most practitioners assume a first order rate law to describe the rate of attenuation in concentration over time. If the rate of attenuation of MTBE in ground water in a monitoring well is controlled by the rate of physical weathering of MTBE from residual gasoline in the source area of a plume, the rate of weathering is constrained by mass transfer limitations on dissolution of the MTBE from gasoline such as diffusion from regions of low hydraulic conductivity to regions of higher conductivity. As a consequence, the instantaneous rate of weathering should be proportional to the amount of MTBE in the residual gasoline, and the rate of attenuation over time should be a first order process.

If the time allowed to reach a specific cleanup level for MTBE has been determined, and if attenuation follows a first order law, the rate of attenuation necessary to meet the goal ($k_{necessary}$) can be calculated from Equation (7.1), where C_g is the cleanup goal, C_o is the current concentration, and t is the time allowed for meeting the goal.

$$k_{necessary} = -\ln\left(C_g \middle/ C_o\right) \middle/ t \qquad \text{Equation (7.1)}$$

In Equation (7.1), $k_{necessary}$ is defined as a rate of attenuation. It has a positive value when concentrations are decreasing over time.

If risk is being managed at the site, but the time to reach a cleanup goal has not been determined, then C_g is any value less than C_o, and $k_{necessary}$ is any value greater than zero.

The achieved rate of attenuation over a time interval being evaluated ($k_{achieved}$) will be defined as the best estimate of the rate of attenuation that is extracted by statistical analysis of the monitoring data. The achieved rate is best determined as the slope of a regression of the natural logarithm of the concentration on time. In the following material in this section, a spreadsheet will be used to extract $k_{achieved}$ from monitoring data. The spreadsheet will also be used to extract statistical confidence intervals on $k_{achieved}$. The slower confidence interval ($k_{with-confidence}$) can be compared to zero to determine if the achieved rate of attenuation is statistically significant. A spreadsheet will also be used to extract the slowest rate of attenuation (k_{detect}) that is statistically different from zero at the predetermined level of confidence.

As will be discussed later, the slower confidence interval is not the "lower" confidence interval identified in the spreadsheet. The spreadsheet calculates a rate of change, not a rate of attenuation. If concentrations are attenuating, the rate of change is negative. If concentrations are attenuating more rapidly, the rate constant is more negative. A "lower" confidence interval as identified by the spreadsheet will be the most negative confidence interval. The "lower" confidence interval will actually be the faster confidence interval, and the "higher" confidence interval will be the slower confidence interval.

The spreadsheet uses the t statistic to calculate confidence intervals on the rate of attenuation. Use of the t statistic requires an assumption that the variance of the data is independent of the values of the data. Variance is a statistical definition of the variation in sample data about a calculated summary statistic. It is a numerical measure of the scatter in the data.

The concentrations of contaminants in ground water generally do not meet the assumption that the variance is independent of the values of the data. The variance in ground water data tends to be proportional to the concentration. Large concentrations have high variance, and small concentrations have a smaller variance. However, the variance of the logarithms of the concentrations is much less dependent on the concentration. To adjust the variance between the higher concentrations in the early samples and the lower concentrations in the later samples, the statistical comparisons will be made between the natural logarithms of the concentrations. This relationship is illustrated in Figure 7.1.

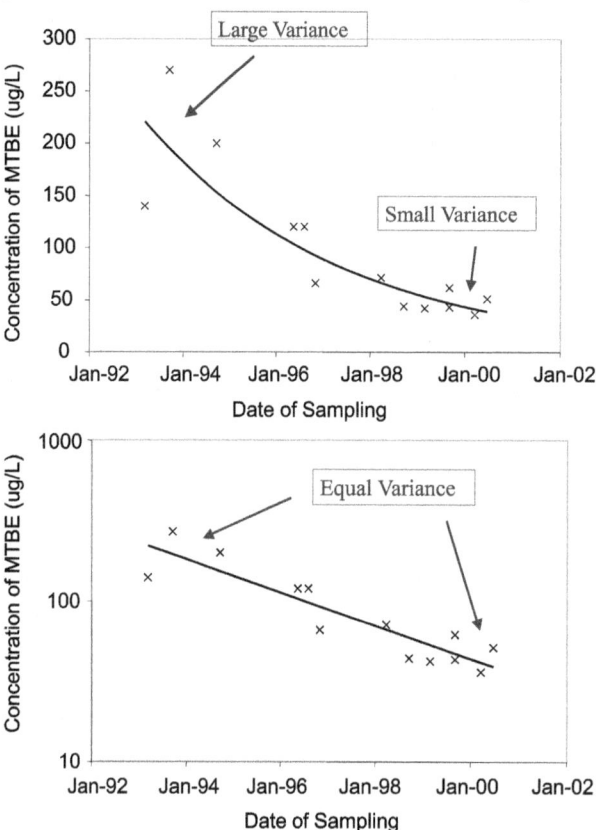

Figure 7.1 *The variance in monitoring data is often proportional to the concentration. The upper panel plots concentration on an arithmetic scale, while the lower panel plots concentration on a logarithmic scale. The variance of the logarithm of the concentration is less dependent on the concentration.*

Table 7.1 compares the relationship between the rates of attenuation necessary to achieve a goal for cleanup, and the achieved rates of attenuation in particular wells. The process of extracting the rate constants from monitoring data will be illustrated with monitoring data from a gasoline spill site in Parsippany, New Jersey. Natural attenuation of MTBE in ground water in this plume is dominated by natural anaerobic biodegradation (Kolhatkar et al., 2002). The locations of the monitoring wells are depicted in Figure 7.2. A portion of the long-term monitoring record is presented in Table 7.2.

Table 7.1 Relationship Between the Rate of Attenuation Necessary for Risk Management or for Monitored Natural Attenuation, and the Achieved Rates of Attenuation During Long-term Monitoring

Remediation Objective a Shrinking Plume	Interpretation
$k_{\text{with-confidence}} > 0$	At the predetermined level of confidence, the concentrations are attenuating over time.
$k_{\text{achieved}} > 0 > k_{\text{with-confidence}}$	Concentrations might be attenuating over time, but there is no statistical confidence that concentrations are attenuating.
Remediation Objective a Cleanup Goal.	
$k_{\text{with-confidence}} > k_{\text{necessary}}$	At the predetermined level of confidence, the concentration goal should be achieved in the specified time.
$k_{\text{achieved}} > k_{\text{necessary}}$ but $k_{\text{with-confidence}} < k_{\text{necessary}}$	Attenuation might achieve the goal in the specified time, but there is no statistical confidence that the rate is adequate.
$k_{\text{achieved}} > k_{\text{detect}}$ but $k_{\text{achieved}} < k_{\text{necesary}}$	Attenuation is happening, but it may not be rapid enough to reach the goal in the specified time.
$k_{\text{necessary}} < k_{\text{detect}}$	Data are too variable or too few to determine if attenuation is proceeding at a rate necessary to meet the goal in the specified time.
$k_{\text{achieved}} < k_{\text{detect}}$	The data are too variable or too few to determine if attenuation is occurring over time.

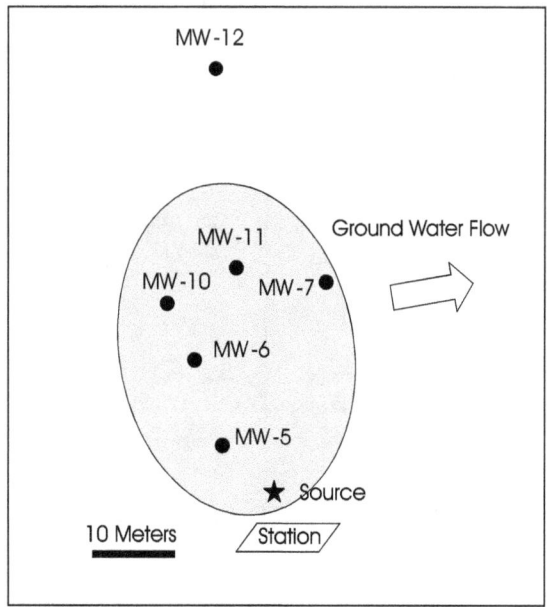

Figure 7.2 *Location of monitoring wells in a plume of MTBE at Parsippany, New Jersey.*

Table 7.2 Long-term Monitoring Data at a Gasoline Spill Site at Parsippany, New Jersey

Date	MW-5	MW-11	MW-6	MW-10	MW-7
			μg/L		
12-Mar-1993	1,500	---	140	---	19*
11-May-1993	---	---	--	290	--
17-Sep-1993	1,900	---	270	180	30*
23-Sep-1994	1,800	2,200	200	5.3	150
17-May-1996	1,300	880	120	5.3	100
10-Aug-1996	980	1.1*	120	23	20
7-Nov-1996	620	660	66	13	17
8-Dec-1997	500	339	--	--	--
27-Mar-1998	635	426	71.2	3.4	--
23-Jul-1998	470	419	--	--	--
18-Sep-1998	1,210	---	44	ND*	--
16-Dec-1998	379	144	--	--	--
1-Mar-1999	700	123	42.2	4.41	--
21-Jun-1999	574	464	---	---	---
7-Sep-1999	1,050	155	43.2	16	---
30-Dec-1999	525	220	---	---	---
20-Mar-2000	501	173	36	6.4	---
22-Jun-2000	420	146	140	5.2	3.7

Not included in the regression.

The following illustrates the process to extract the rate constants from monitoring data. Microsoft EXCEL will be used in the illustration because it is widely available to regulators in state agencies. The data are from MW-5 in Table 7.2. Despite the best efforts at quality control, any large data set contains spurious data. Professional judgment was used to exclude selected data in Table 7.2 from the regression.

Enter the dates the well was sampled (Column G in the example), and the concentrations of MTBE (Column H in the example). Enter the formula for taking the natural logarithm of the contents of Cell H1 into Cell I1 [=LN(H2)]. Use the mouse to click any cell other than Cell I1, and Excel will accept the formula. After the formula is accepted, drag it through the other cells in Column I to calculate the natural logarithm of all the data.

Then open the Tools menu from the menu bar and select Data Analysis.

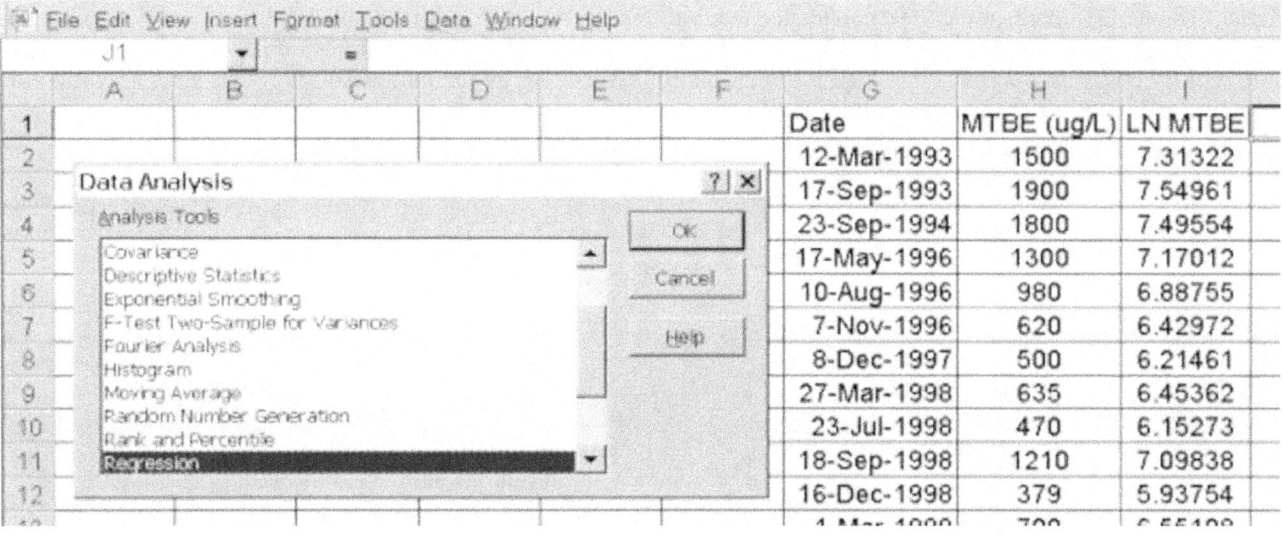

	A	B		E	F	G	H	I	
1						Date	MTBE (ug/L)	LN MTBE	
2						12-Mar-1993	1500	7.31322	
3						17-Sep-1993	1900	7.549609	
4						23-Sep-1994	1800	7.495542	
5						17-May-1996	1300	7.17012	
6						10-Aug-1996	980	6.887553	
7						7-Nov-1996	620	6.429719	
8						8-Dec-1997	500	6.214608	
9						27-Mar-1998	635	6.453625	
10						23-Jul-1998	470	6.152733	
11						18-Sep-1998	1210	7.098376	
12						16-Dec-1998	379	5.937536	
13						1-Mar-1999	700	6.55108	
14						21-Jun-1999	574	6.352629	
15						7-Sep-1999	1050	6.956545	
16						30-Dec-1999	525	6.263398	
17						20-Mar-2000	501	6.216606	
18						22-Jun-2000	420	6.040255	

Open the Data Analysis menu and select Regression. -

	A	B	C	D	E	F	G	H	I
1							Date	MTBE (ug/L)	LN MTBE
2							12-Mar-1993	1500	7.31322
3							17-Sep-1993	1900	7.54961
4							23-Sep-1994	1800	7.49554
5							17-May-1996	1300	7.17012
6							10-Aug-1996	980	6.88755
7							7-Nov-1996	620	6.42972
8							8-Dec-1997	500	6.21461
9							27-Mar-1998	635	6.45362
10							23-Jul-1998	470	6.15273
11							18-Sep-1998	1210	7.09838
12							16-Dec-1998	379	5.93754

Data Analysis

Analysis Tools

Covariance
Descriptive Statistics
Exponential Smoothing
F-Test Two-Sample for Variances
Fourier Analysis
Histogram
Moving Average
Random Number Generation
Rank and Percentile
Regression

OK
Cancel
Help

Perform a linear regression of the natural logarithm of the concentrations of MTBE on the date the water samples were collected from monitoring well #5.

Cell locations of the calculated natural logarithms of the concentrations are entered into the Input Y Range window, and the dates are entered in the Input X Range window. The cell names can be typed into the windows, or click with the mouse on the window, erase any names already entered, then select the data to be entered with the mouse. If you want a level of statistical confidence that is different than 95% confidence, enter the desired level in the appropriate window. In the example, a confidence level of 80% was selected. Click OK to perform the regression.

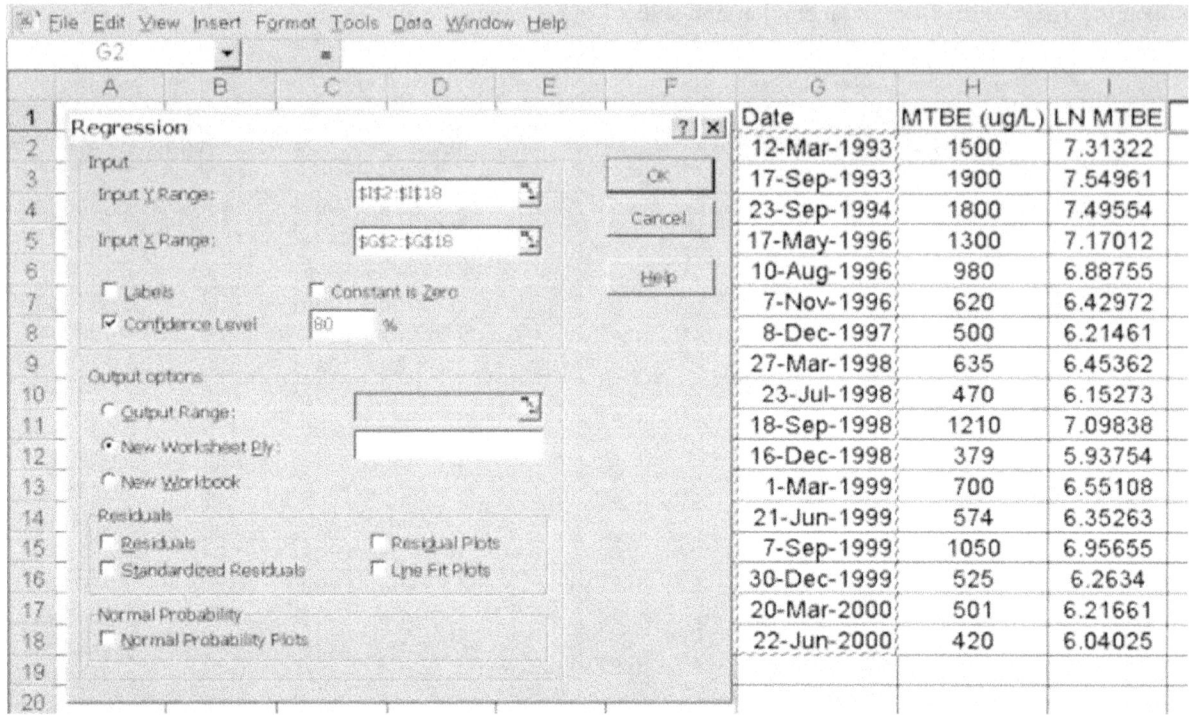

Date	MTBE (ug/L)	LN MTBE
12-Mar-1993	1500	7.31322
17-Sep-1993	1900	7.54961
23-Sep-1994	1800	7.49554
17-May-1996	1300	7.17012
10-Aug-1996	980	6.88755
7-Nov-1996	620	6.42972
8-Dec-1997	500	6.21461
27-Mar-1998	635	6.45362
23-Jul-1998	470	6.15273
18-Sep-1998	1210	7.09838
16-Dec-1998	379	5.93754
1-Mar-1999	700	6.55108
21-Jun-1999	574	6.35263
7-Sep-1999	1050	6.95655
30-Dec-1999	525	6.2634
20-Mar-2000	501	6.21661
22-Jun-2000	420	6.04025

Excel returns a SUMMARY OUTPUT of the regression as a new tab in the workbook. The first order rate of change of concentration with time is located in cell B18. This is the negative of the rate of attenuation. Because Excel was supplied a date as the X variable, the rate constant is reported in units of day^{-1}.

Cells F18 and G18 contain the 95% confidence interval on the rate, and cells H18 and I18 have the 80% confidence interval on the rate. The rate of change and the confidence intervals were converted from units of *per day* to units of *per year* by inserting a formula into Cell B19 that multiplied the rate by the number of days in a year, accepting the formula, and dragging the formula onto cells F19 through I19.

B19 =B18*365.25

	A	B	C	D	E	F	G	H	I
1	SUMMARY OUTPUT								
2									
3	*Regression Statistics*								
4	Multiple R	0.77612007							
5	R Square	0.60236237							
6	Adjusted R	0.57585319							
7	Standard E	0.34248874							
8	Observatio	17							
9									
10	ANOVA								
11		*df*	*SS*	*MS*	*F*	*Significance F*			
12	Regressior	1	2.665349815	2.66535	22.72279	0.00024962			
13	Residual	15	1.759478077	0.117299					
14	Total	16	4.424827892						
15									
16		*Coefficients*	*Standard Error*	*t Stat*	*P-value*	*Lower 95%*	*Upper 95%*	*Lower 80.0%*	*Upper 80.0%*
17	Intercept	24.4325875	3.730990758	6.548552	9.22E-06	16.480164	32.38501093	19.430801	29.43437392
18	X Variable	-0.0004977	0.000104402	-4.76684	0.00025	-0.0007202	-0.00027514	-0.00063763	-0.000357705
19	per year	-0.1817725				-0.2630504	-0.10049455	-0.23289335	-0.1306516
20									

The confidence intervals on the rate of change are calculated in Excel using the Student's t distribution as defined in Equation 7.2, where $-k$ is the statistic of interest (the slope of the regression line), and s_k is the standard deviation or standard error of $-k$.

$$t = \left| -k \middle/ s_k \right|$$

Equation 7.2

Think of the t statistic as the ratio of signal to noise. If the data used to calculate the t statistic have a normal distribution, the values of t are known for any level of confidence.

Excel calculates the confidence interval on $-k$ following Equation 7.3:

$$\text{confidence interval on } -k = -k - t * s_k$$

Equation 7.3

The probability of error (α) is the probability that a calculated rate constant will be accepted as a statistically significant rate, even though the calculated rate was a result of random variation and was not truly different from zero. The confidence level is one minus the probability of error. At a confidence level of 90%, the probability of error is 10%. Excel calculates the t statistic using a two-tailed distribution of errors in the estimate of the rate of change. Half of the error in α is associated with rates where concentrations are increasing over time, and half is associated with rates of attenuation. We are only interested in the half of the error that is associated with rates of attenuation. The 90% confidence intervals calculated by Excel on the rate of change are 95% confidence intervals on the rate of attenuation. The 95% confidence intervals on the rate of change are 97.5% confidence intervals on the rate of attenuation.

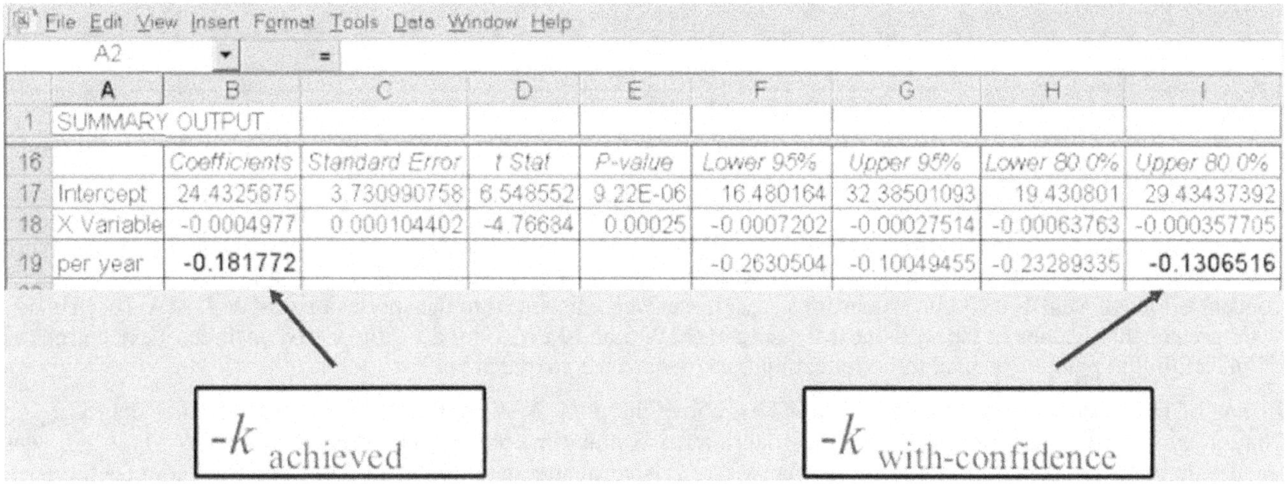

In the screen shot above from output of the linear regression, the rate of change in concentration with time is -0.181772 per year (Cell B19), corresponding to a rate of attenuation $k_{achieved}$ of 0.18 per year. The *Upper 80.0%* Confidence interval on the rate of change is -0.1306516 per year (Cell G19), which corresponds to a $k_{with-confidence}$ on the rate of attenuation of 0.13 per year at the 90% confidence level.

The difference between $k_{achieved}$ and $k_{with-confidence}$ is the value that a calculated $k_{achieved}$ must exceed to be statistically different from zero (Equation 7.4). This difference is the minimum rate of attenuation that can be detected at the accepted level of confidence (k_{detect}) with the existing variability in the data. In this case, the value of k_{detect} is 0.181772 per year minus 0.1306516 per year or 0.051 per year.

$$k_{achieved} - k_{with-confidence} = k_{detect}$$

Equation 7.4

Table 7.3 compares the concentrations of MTBE and the rates of attenuation of MTBE in all five wells in the plume at Parsippany, New Jersey (see Table 7.2 and Figure 7.2). Table 7.3 summarizes an analysis of monitoring data from March 1993 through June 2000. The current goal for MTBE in New Jersey is 70 μg/L. To illustrate comparisons of the rate constants, we will assume that a hypothetical goal of 20 μg/L should be reached in a "reasonable" interval of five years. The rate of attenuation necessary to meet the goal ($k_{necessary}$) was calculated using Equation 7.1.

Table 7.3   Progress of Natural Attenuation of MTBE at a Gasoline Spill Site at Parsippany, New Jersey

Well	MW-5	MW-11	MW-6	MW-10	MW-7
20 µg/L is the hypothetical goal to close the site chosen for this illustration.					
5.0 years is the hypothetical "reasonable" time period to reach the goal for this site, chosen for this illustration.					
Current Concentration (µg/L)	420	146	51	5.2	3.7
Maximum Concentration (µg/L)	1,900	2,200	270	290	150
$k_{necessary}$ (per year)	0.61	0.39	0.19	Already met goal	Already met goal
$k_{achieved}$ (per year)	0.18	0.45	0.27	0.41	0.64
$k_{with\text{-}confidence}$ (per year) 95% confidence	0.13	0.36	0.22	0.21	0.36
k_{detect} (per year)	0.05	0.09	0.05	0.20	0.28
number of sampling dates	17	13	11	11	5

Initially, five wells had concentrations of MTBE that were higher than 20 µg/L. By 2000, two of the five wells had reached the cleanup goal. These wells (MW-10 and MW-7) were at the lateral margins of the plume (compare Figure 7.2). Two wells near the source area (MW-5 and MW-6) and one distant well (MW-11) still maintained significant concentrations of MTBE in 2000. Values for $k_{achieved}$ in the wells distant from the source area (MW-7, MW-10, MW-11) were greater than values in the wells near the source (MW-5 and MW-6) by a factor of two or three. Concentrations of MTBE in the plume appeared to be retreating back toward the source area.

In one of the wells with concentrations of MTBE above the goal (MW-6), the $k_{with\text{-}confidence}$ was greater than $k_{necesssary}$. Natural attenuation was on track to meet the goal in a reasonable time period. In a second well (MW-11), $k_{achieved}$ was greater than $k_{necessary}$, but $k_{with\text{-}confidence}$ was less than $k_{necessary}$. Natural attenuation may have been on track to meet the goal, but the data were too variable or too few to support the projection at a 90% confidence level. In the well with the highest concentration of MTBE (MW-5), $k_{achieved}$ was much less than $k_{necessary}$. In order to meet the goal in five years, it would be necessary to actively remediate the source area near MW-5.

Section 8 -

Typical Rates of Attenuation in Source Areas -

Regulators are often asked to determine the number of rounds of sampling that is necessary for them to evaluate the behavior of an MTBE plume. To plan a monitoring effort, it is necessary to have some idea of the rate of attenuation MTBE in the source area of a plume and the variation in that rate over time. This section illustrates the range of rates that might be expected at a typical gasoline spill site. At a major proportion of MTBE sites, the long-term monitoring data will fail to show that MTBE is attenuating in the most contaminated wells. It also illustrates the number of rounds of sampling that are necessary to document the rate of natural attenuation with statistical significance. In general, short data sets with less than twelve samples may fail to detect rates of attenuation that have environmental significance.

8.1   Typical Rates of Attenuation Over Time in Source Areas

Wilson and Kolhatkar (2002) compared the rate of attenuation of MTBE over time in the source area of five plumes to the rate of attenuation of MTBE in ground water along the flow path. The rates of attenuation in ground water were from two-fold to more than ten-fold faster. Thus, in the source area, the persistence of the plumes is controlled by the rate of attenuation of concentrations in the source areas. Durrant et al., (1999) explained the long-term persistence of a plume of MTBE in California by modeling the diffusion of high concentrations of MTBE into regions of low hydraulic conductivity early in the spill. Over time, the MTBE in the residual gasoline dissolved into ground water and was carried away by the flow of ground water. After MTBE in the residual gasoline was depleted, the plume was sustained by the slow diffusion of MTBE back out of the regions of low hydraulic permeability.

Other situations can produce the same behavior. If residual gasoline remains in the aquifer, MTBE can slowly partition from the residual gasoline to the ground water. Peargin (2000, 2001) compared the relative rate of attenuation of MTBE, benzene, and xylenes in wells in the smear zones of gasoline spills. The rate of attenuation was independent of the water solubility of the contaminant and could not be explained by the expected rate of dissolution from gasoline into water. Peargin (2000, 2001) concluded that mass transfer limitations slowed the transfer of MTBE from the gasoline to the active flow paths in the aquifer.

Wilson and Kolhatkar (2002) extracted the rate of attenuation of MTBE in the most contaminated well at gasoline spills in California, Florida, North Carolina, New York, and the site at Parsippany, New Jersey, that was discussed extensively in the previous section. The rate of attenuation at the five sites varied from 0.15 per year to 0.75 per year. The rate that was statistically significant at 90% confidence varied from 0.04 per year to 0.29 per year. Peargin (2000, 2001) extracted the rate of natural attenuation of MTBE in 23 wells in the smear zone of 15 gasoline stations in the eastern U.S. (primarily Maryland). The fastest rate of attenuation of the source was 0.7 per year, equivalent to a half-life of one year. The mean rate of attenuation was 0.04 per year, equivalent to a half-life of 17 years. Shorr and Rifai (2002) calculated the rate of change in the concentration of MTBE over time for 694 monitoring wells at gasoline spill sites in Texas. In two thirds of the wells, the concentrations of MTBE declined over time. In the wells where the concentrations of MTBE declined over time, the median rate of attenuation was 0.043 per year, corresponding to a half-life of 16 years, and 25% had a rate equal to, or greater than, 0.37 per year, equivalent to a half-life near two years. Robb and Moyer (2003) provided monitoring data on a site in the Midwestern U.S. The rate of attenuation of MTBE in the most contaminated well was 0.62 per year (0.29 per year at 90% confidence) equal to a half-life of 1.1 years.

Figure 8.1 collates the rates of attenuation of MTBE over time in the most contaminated well at thirteen of the sites described by Peargin (2000, 2001) that had not been subjected to remediation, at five sites discussed in Wilson and Kolhatkar (2002), at the site in the Midwestern U.S. discussed by Robb and Moyer (2003), and at the site in South Carolina described in Landmeyer et al., (1998). Landmeyer (personal communication J. Landmeyer, USGS, Columbia, SC) provided monitoring data on attenuation of MTBE in the most contaminated well at the site in South Carolina.

Twenty sites are not a statistically representative sample of the hundreds of thousands of MTBE sites in the United States; however, the distribution of the rates of attenuation can at least illustrate the possible behavior of MTBE in gasoline spill sites. In six of twenty sites, the concentration of MTBE increased over time in the most contaminated well, instead of

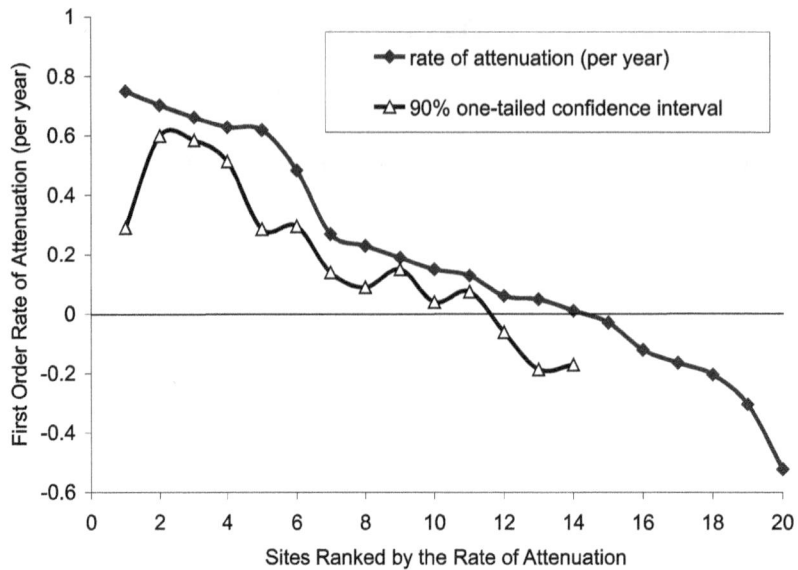

Figure 8.1 -*Distribution of the rates of attenuation of MTBE over time in source areas of plumes from gasoline spills. Negative rates indicate increasing concentrations with time.*

attenuating. At 14 sites, concentrations of MTBE attenuated over time. At 11 of the sites, the 90% one-tailed confidence interval on the rate was greater than zero. At a major proportion of MTBE sites, long-term monitoring data will fail to show that MTBE is attenuating in the most contaminated wells, or will fail to show that MTBE is attenuating at 90% confidence. If the rate of attenuation is truly greater than zero, but the rate of attenuation is slow and variable, the chance that a calculated rate of attenuation will be found to be statistically different from zero is strongly influenced by the number of samples used to calculate the rate. As the number of samples goes up, the proportion of calculated rates that are found to be statistically greater than zero will also increase. To illustrate this effect, Figure 8.2 compares the minimum rate of attenuation that was detectable at 90% confidence to the number of samples used to extract the rates presented in Figure 8.1.

8.2 Number of Sampling Dates Needed to Calculate Rates of Attenuation

A regression analysis of long-term trends in monitoring data is subject to two kinds of error. The analysis may fail to detect attenuation when the attenuation is really happening. This happens most often when the rate of attenuation is slow and variable, and there are simply not enough data to distinguish the true trend above the natural variation in the data. If this is the case, the chance that a calculated rate of attenuation will be found to be statistically different from zero is strongly influenced by the number of sampling dates used to calculate the rate. As the number of sampling dates increases, the proportion of calculated rates that are found to be statistically greater than zero will also increase.

Often state agencies only have sampling data for a limited number of dates. To determine the minimum number of sample dates that are needed to extract a rate that is statistically significant, a regression analysis was performed on small portions of the long-term monitoring data used to extract the rates in Figure 8.1. Portions of each long-term record were selected that contained four, five, six, or more dates. To avoid bias, the dates in the portions of the record were selected to distribute the dates in the portion equally about the central date of the parent record. Then the portions were analyzed as described in Section 7 to determine whether the rate of attenuation was statistically different from zero. The number of dates in the portion were expanded until regression analysis indicated that rate of attenuation was statistically significant with a one-tailed confidence level of 90%. Then the number of dates was expanded again until the rate was significant at 95% confidence.

The number of sampling dates required for the rate to be statistically significant is presented in Figure 8.2. For many of the data sets, as few as four samples were adequate to extract a rate statistically greater than zero. Several state agencies will evaluate a site for natural attenuation after two years of quarterly monitoring. Eight quarters of monitoring would have failed to recognize natural attenuation at 6 of the 14 sites. However, eight sampling dates are an efficient size for the minimum data set to evaluate natural attenuation. Eight sampling dates were sufficient to recognize roughly one-half of the sites where natural attenuation was occurring.

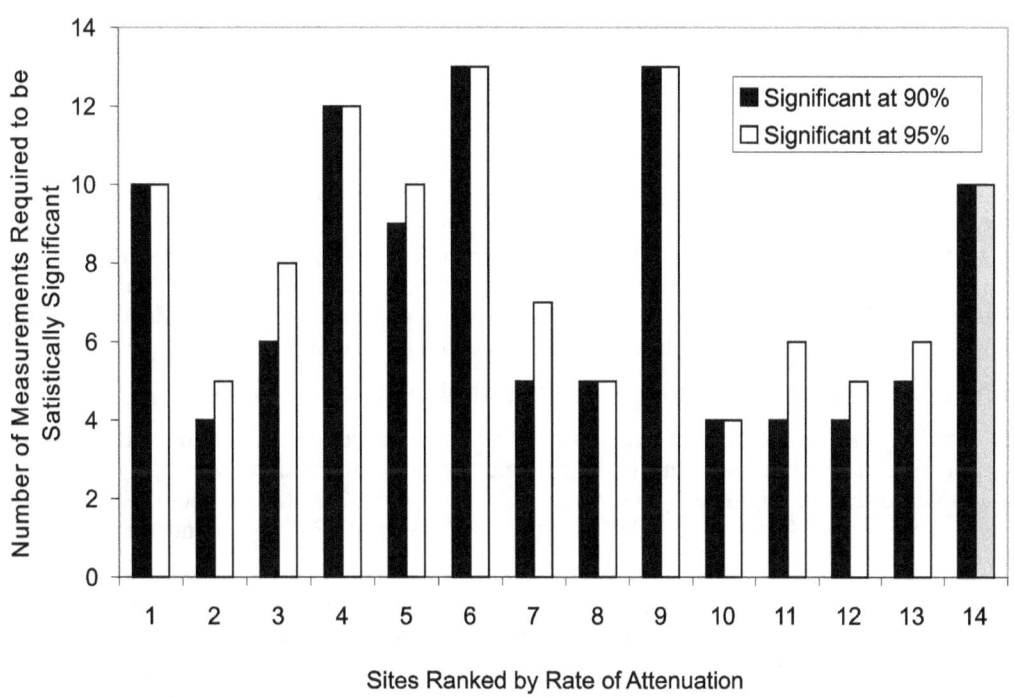

Figure 8.2. *Variation in the number of sampling dates in a data set required to extract a rate of natural attenuation that is statistically significant.*

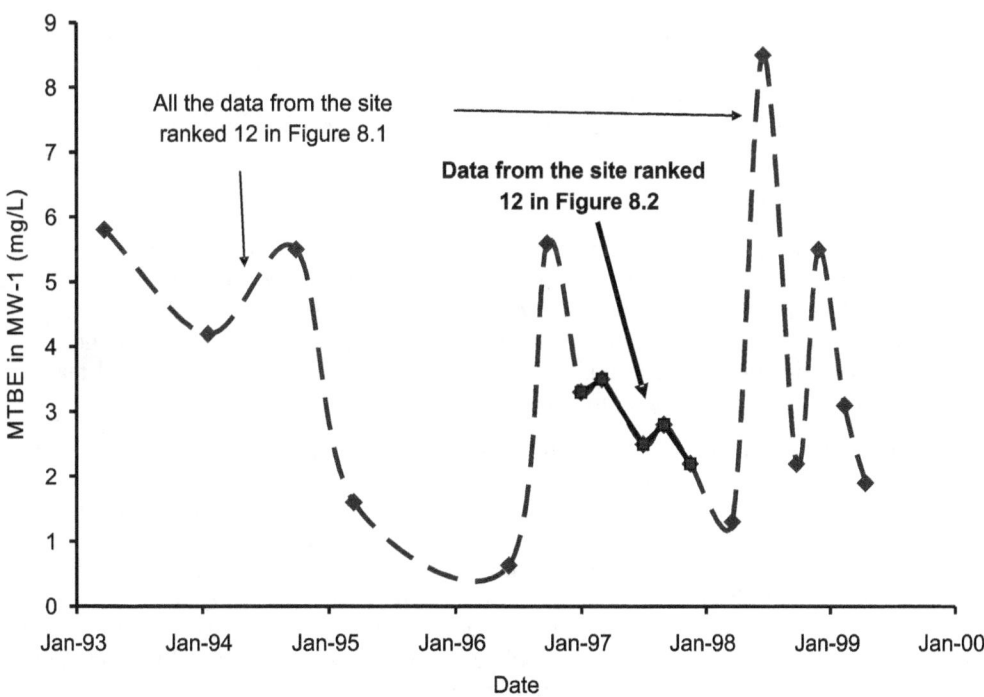

Figure 8.3. *Potential for error when a short data set is used to estimate the rate of attenuation of concentrations over time.*

The level of statistical confidence in the regression analysis had little effect on the number of sampling dates required. There was little difference between the number of sample dates needed to extract a rate that was significant at 90% confidence and a rate that was significant at 95% confidence.

In the second kind of error, a trend that is recognized as significant over a few dates is not borne out over the longer interval, because the shorter interval used to evaluate the rate of attenuation was not representative of the longer interval. This effect is illustrated in Figure 8.3, using data from the site ranked number 12 in Figures 8.1 and 8.2. The rate of attenuation calculated for the entire data set as presented in Figure 8.1 was 0.06 per year, which was not significantly different from zero at 90% confidence. The rate calculated for the five central dates as presented in Figure 8.3 is 0.49 per year. These five dates were used to calculate the minimum number of dates required in Figure 8.2. Over the smaller data set, the rate of attenuation was much faster, and the rate was significant at 95% confidence. The only protection from this second kind of error is more sampling dates that extend over a longer period of time.

8.3 Effect of Number of Sampling Dates on the Detectable Rate of Attenuation

The number of sampling dates in the data set also has an effect on the detection limit for the rate of natural attenuation. As discussed in Section 7, the detection limit is the minimum rate of attenuation statistically different from zero at some level of confidence (k_{detect}). Figure 8.4 presents the relationship between the number of dates in the data set and k_{detect} at 90% confidence for the 14 sites presented in Figure 8.1 where MTBE was attenuating over time in the most contaminated well.

Most of the data sets from the 14 sites are more extensive than is usually available. To estimate minimum detectable rates that would be extracted with fewer data, Figure 8.4 also presents the minimum rates of attenuation that would be detected when the rates were calculated with half of the available data. The rates were calculated from half of the data that occupied the middle portion of the monitoring record. The first portion (approximately 25%) and final portion (approximately 25%) of the monitoring data were excluded.

In their survey of the rate of attenuation of MTBE at sites in Texas, Shorr and Rifai (2002) found the rate of attenuation met or exceeded 0.001 per day, or 0.37 per year in only 17% of wells. As discussed in Section 2, the median concentration of MTBE in the most contaminated well at gasoline spill sites in Texas is near 1,000 μg/L. If the rate of natural attenuation is 0.37 per year, a site with a maximum concentration of 1,000 μg/L of MTBE would require 12 years to reach the EPA advisory concentration of 20 μg/L.

A data set to evaluate natural attenuation of MTBE should have a detection limit (k_{detect}) lower than 0.37 per year. When the number of sampling dates in Figure 8.4 was greater than 12, the detection limit for attenuation was less than 0.37 per year. When the number of sampling dates was less than 12, the detection limit for most of the data sets was greater than 0.37 per year. Short data sets with less than 12 samples may fail to detect rates of attenuation that have environmental significance.

8.4 Effect of Seasonal Variations

The statistical considerations discussed so far ignore any seasonal effects on the concentration of contaminants. If there are strong seasonal effects on the recharge of precipitation to ground water, these effects may be reflected in the measured concentrations of MTBE in monitoring wells. As the water table moves up and down in response to recharge, the water may wet more or less of the gasoline in the smear zone, resulting in higher or lower concentrations of MTBE in the well. If a plume is vertically heterogeneous, the screened interval of a monitoring well may sample different vertical regions in the same plume as the water table shifts.

Seasonal variations can add to the variability in short-term data sets. When there are strong seasonal effects, it may be useful to extract the rate of attenuation of the seasonal maximum concentrations, and compare that rate to the overall rate of attenuation. Figure 8.5 illustrates a data set from a well with a strong seasonal component. The concentration maximums are associated with the summer months in 1993, 1994, 1996, and 1998. However, not every summer shows a maximum. There was no maximum in the summer of 1995, and the well was not sampled in the summer of 1997. In this case, there was no difference in the rate of attenuation of the seasonal maximum concentrations and the concentrations throughout the year. When the rate of attenuation is extracted from the complete data set (connected by the solid line in Figure 8.5), the rate of attenuation is 0.48 per year, or 0.30 per year at 90% confidence. If the rate is extracted from the four summer maxima (dashed line in Figure 8.3), the rate is 0.47 per year, or 0.34 per year at 90% confidence.

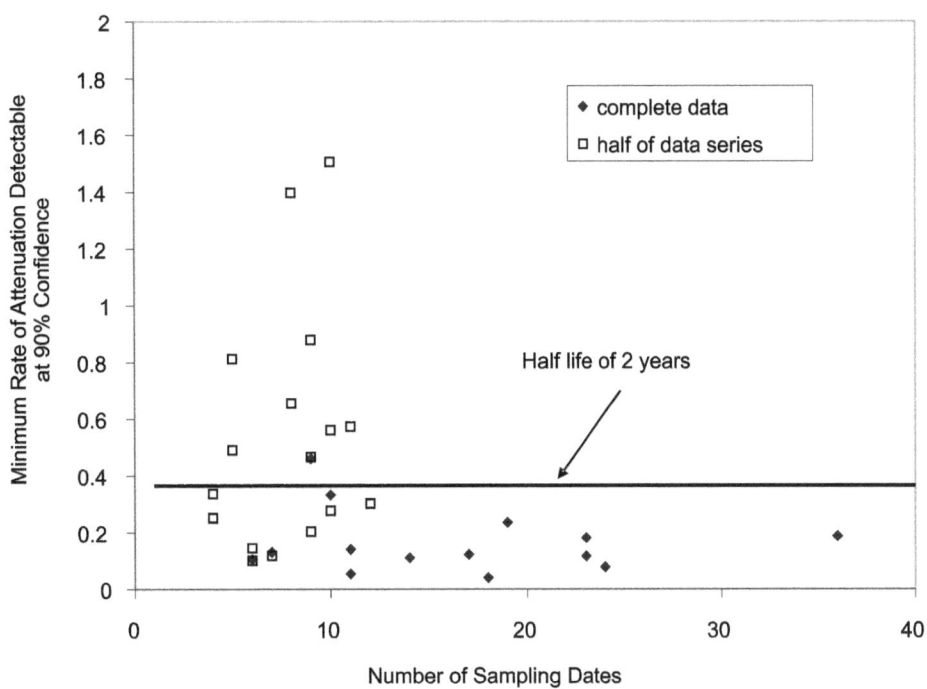

Figure 8.4 *Effect of the number of samples used to calculate a rate of attenuation on the minimum rate of attenuation that is statistically different from zero at 90% confidence.*

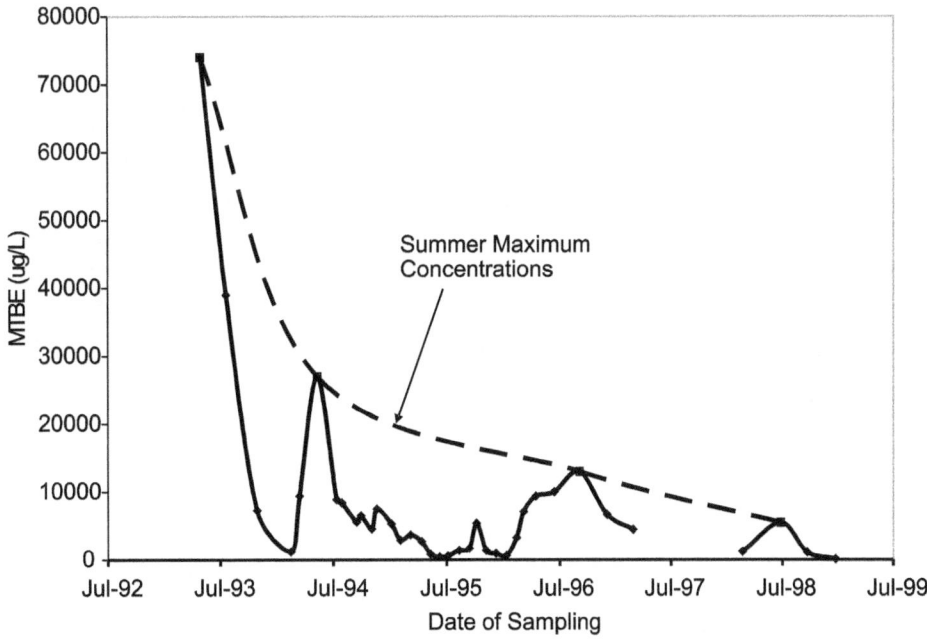

Figure 8.5 *A monitoring record from a well in Maryland with seasonal maximum concentrations of MTBE in certain years.*

Section 9 -

Quality Assurance Statement -

9.1 Analysis of Concentrations in Water

Laboratory analyses for data presented in Table 4.3, Figure 4.4, and Figure 4.5 were conducted at the Robert S. Kerr Environmental Research Center in accordance with a Quality Assurance Project Plan prepared for in-house task 10013 (Fate of Fuel Oxygenates in Aquifer Materials, approved January 2001). Concentrations of MTBE, TBA, and benzene were determined following in-house SOPs very similar to Lin et al., 2003. Water samples were prepared with a heated static headspace sampler, and determined by gas chromatography and mass spectrometry. Ethanol in water was determined by direct aqueous injection onto a gas chromatograph equipped with a flame ionization detector.

The stated data quality objectives for analysis of MTBE, TBA, benzene, and ethanol, were as follows: The reported concentration of continuing calibration check standards and matrix spikes will agree with the expected concentration plus or minus 15% of the known concentration. Analytical duplicates will agree with each other plus or minus 15%.

The microcosms constructed with sediment from the gasoline spill site at Parsippany, New Jersey, were the only microcosms that acclimated and degraded MTBE under anaerobic conditions. Tables 9.1 and 9.2 summarize typical data quality for MTBE in the microcosms constructed with material from Parsippany, New Jersey. The first five sampling dates correspond to the data presented in Figure 4.4. Five out of 22 of the calibration check standards did not meet the goal of ± 15% of the nominal value (Table 9.1). However, none of the check standards varied by more 25% from the nominal value. Two out of nine laboratory duplicates did not agree within 15%, but all the duplicates agreed within a relative percent difference of 25% (Table 9.2). Two out of six of the matrix spikes exceed the goal of ± 15%, but all the matrix spikes were within ± 25% (Table 9.2).

The method blanks were always less than 0.5 µg/L with the exception of samples collected on 11/16/01 (Table 9.2). On this date, the analysts diluted all the samples provided to him by two fold, to have enough material for a laboratory duplicate.

One of the treatments included in the experiment was a container control, containing the analytes of interest in sterile water. At most sampling intervals, triplicate container controls were analyzed. The variation between the triplicate analyses represents variability in the construction of the microcosms as well as any error in the analysis of the concentrations of the analyte. The sample standard deviation was never more than 5% of the mean of the triplicate analyses (Table 9.2). The variation between the mean of triplicate samples in the "living" microcosms was always much higher (See Figure 4.4), reflecting the influence of sorption and biodegradation on concentrations of the analytes.

Due to unavoidable problems with laboratory infrastructure, some of the samples were held for up to 116 days before analysis. The samples were preserved in trisodium phosphate. There was no indication from the container controls that the long holding time caused any loss of MTBE (Table 9.2). There was no statistically significant difference between the mean concentration of MTBE at the beginning of the experiment and the concentration after 358 days of incubation. The mean concentration of MTBE in the container controls varied as much as 9% from one sampling time to the next (6/28/01 to 7/26/01). Variation in mean concentrations in the container controls most likely reflected normal variations in the calibration of the analytical instrument, and not loss or gain of MTBE in the microcosms.

All the data for MTBE were determined to be of acceptable quality, and the data were used in the report.

Tables 9.3 and 9.4 summarize typical data quality for TBA in the microcosms constructed with material from Parsippany, New Jersey. Analyses of TBA were not as accurate or precise as the analyses of MTBE. Six out of 22 of the calibration check standards did not meet the goal of ± 15% of the nominal value (Table 9.3). Two of the six did not meet a goal of ± 25% of the nominal value. One check sample was reported as 155% of the nominal value. Two of seven matrix spike samples did not meet the goal of ± 15% of the spiked value (Table 9.4). One of the matrix spike samples reported 126% of the spiked concentration.

The method blanks were always less than 10 µg/L with the exception of samples collected on 11/16/01. On this date, the analysts diluted all the samples provided to him by two fold, to have enough material for a laboratory duplicate.

The standard deviation of the samples in the container controls was 10% or less of the mean. The mean concentration of MTBE in the container controls varied as much as 36% from one sampling time to the next (5/2/01 to 10/25/01).

All the data for TBA were determined to be of acceptable quality, and the data were used in the report.

Tables 9.5 and 9.6 summarize typical data quality for benzene in the microcosms constructed with material from Parsippany, New Jersey. Analyses of benzene were not as accurate and precise as the analyses of MTBE. Seven out of 22 of the calibration check standards did not meet the goal of ± 15% of the nominal value (Table 9.5). However, all of them did meet a goal of ± 25% of the nominal value. One check sample was reported as 123% of the nominal value. Two of seven matrix spike samples did not meet the goal of ±15% of the spiked value (Table 9.6). However, all of the matrix spike samples met a goal of ± 25% of the spiked concentration.

The method blanks were always less than 0.5 µg/L with the exception of samples collected on 11/16/01. On this date, the analysts diluted all the samples provided to him by two fold, to have enough material for a laboratory duplicate.

The standard deviation of the samples in the container controls was 10% or less of the mean. The mean concentration of benzene in the container controls varied as much as 53% from one sampling time to the next (5/2/01 to 10/25/01)

All the data for benzene were determined to be of acceptable quality, and the data were used in the report.

Tables 9.7 and 9.8 summarize typical data quality for ethanol in the microcosms constructed with material from Parsippany, New Jersey. The continuing calibration standards, duplicates, and matrix spike samples were all within the goal of ± 15%. Ethanol was not included in the container controls. All the blanks were less than 1 mg/L. All the data for ethanol were determined to be of acceptable quality, and the data were used in the report.

Several Tables and Figures in this report reference data provided by state agencies. To our knowledge, these analyses were conducted following EPA 8260 or 8260B (purge and trap with gas chromatography with a mass spectrometer detector). The results should be comparable to results obtained at the Kerr Center.

9.2 Stable Carbon Isotope Analyses

The analyses were performed by the University of Oklahoma, acting as a private contractor. The quality assurance data provided with the samples indicated that the sample standard deviation of determination of $\delta^{13}C$ varied from ± 0.1 ‰ to ± 0.18 ‰. The nature of the analysis makes it impossible to do a matrix spike. Each analysis is referenced to a calibration standard.

Table 9.1 Typical Quality Performance Data for Continuing Calibration Check Standards for MTBE in Water. All Values are µg/L Unless Otherwise Indicated

Date Collected	5/2/01	5/31/01	6/28/01	7/26/01	10/25/01	11/16/01	4/25/02	5/15/03
Date Analyzed	5/31/01	9/24/01	9/3/01	9/21/01	11/30/01	11/26/01	4/25/02	5/16/03
Check Standard Nominal	20	200	200	200	200	200	200	20
Check Standard Measured	21.8	209	204	239	204	228	235	19.6
Percent of Check Standard	109.0%	104.5%	102.0%	119.5%	102.0%	114.0%	117.5%	98.0%
Check Standard Nominal	200	20	200	20	20.0	200	20	100
Check Standard Value	220	21.3	237	21.3	20.4	236	18.1	100
Percent of Check Standard	110.0%	106.5%	118.5%	106.5%	102.0%	118.0%	90.5%	100.0%
Check Standard Nominal	200	200	20	200			200	200
Check Standard Measured	204	193	23.7	233			184	210
Percent of Check Standard	102.0%	96.5%	118.5%	116.5%			92.0%	105.0%

Table 9.2 Typical Quality Performance Data for Analysis of MTBE in Water, Including Blanks, Laboratory Duplicates, and Matrix Spikes. All Values are µg/L Unless Otherwise Indicated

Date Collected	5/2/01	5/31/01	6/28/01	7/26/01	10/25/01	11/16/01	4/25/02	5/15/03
Date Analyzed	5/31/01	9/24/01	9/3/01	9/21/01	11/30/01	11/26/01	4/25/02	5/16/03
Blank 1	no report	<0.5	<0.5	<0.5	<0.5	<1	<0.3	<0.1
Blank 2	no report	<0.5	<0.5	<0.5			<0.3	<0.1
Sample Analysis 1	910	20.2	13.3	15.9	3.3	<1	10.3	65
Laboratory Duplicate 1	886	20.6	15	15.6	2.8	<1	8.45	61.9
Relative Percent Difference	2.7%	2.0%	12.0%	1.9%	16.4%		19.7%	4.9%
Sample Analysis 2	159				6.2			
Laboratory Duplicate 2	164				5.6			
Relative Percent Difference	3.1%				10.2%			
Spike Concentration	200	200	200	200	200	200		
Sample Concentration	134	25.6	17.7	14.9	7.4	<1		
Spike Recovery (Percent)	108%	99%	113%	119%	114%	116%		
Time of Incubation (Days)	0	29	57	85	176	198	358	
Holding Time (Days)	29	116	67	57	36	10	0	
Sterile Water Control 1	39.2	37.4	35.8	37.4	36.6		39.9	
Sterile Water Control 2	39.6	37.0	35.4	39.0	38.5		39.0	
Sterile Water Control 3	38.2	37.6	36.4	41.2	38.5		36.1	
Mean	39.0	37.3	35.9	39.2	37.9		38.3	
Standard Deviation	0.72	0.31	0.50	1.91	1.10		1.99	

Table 9.3 Typical Quality Performance Data for Continuing Calibration Check Standards for TBA in Water. All Values are µg/L Unless Otherwise Indicated

Date Collected	5/2/01	5/31/01	6/28/01	7/26/01	10/25/01	11/16/01	4/25/02	5/15/03
Date Analyzed	5/31/01	9/24/01	9/3/01	9/21/01	11/30/01	11/26/01	4/25/02	5/16/03
Check Standard Nominal	20	200	200	200	200	200	200	20
Check Standard Measured	17.5	207	198	216	237	252	223	22.5
Percent of Check Standard	87.5%	103.5%	99.0%	108.0%	118.5%	126.0%	111.5%	112.5%
Check Standard Nominal	200	20	200	20	20.0	200	20	100
Check Standard Value	168	31.1	192	19.5	24.3	241	22.5	98.6
Percent of Check Standard	84.0%	155.5%	96.0%	97.5%	121.5%	120.5%	112.5%	98.6%
Check Standard Nominal	200	200	20	200			200	200
Check Standard Measured	191	186	21.1	185			239	200
Percent of Check Standard	95.5%	93.0%	105.5%	92.5%			119.5%	100.0%

Table 9.4   Typical Quality Performance Data for Analysis of TBA in Water, Including Blanks, Laboratory Duplicates, and Matrix Spikes. All Values are µg/L Unless Otherwise Indicated

Date Collected	5/2/01	5/31/01	6/28/01	7/26/01	10/25/01	11/16/01	4/25/02	5/15/03
Date Analyzed	5/31/01	9/24/01	9/3/01	9/21/01	11/30/01	11/26/01	4/25/02	5/16/03
Blank 1	no report	<10	<10	<10	<10	<20	<2.6	<2.8
Blank 2	no report	<10	<10	<10	<10		<2.6	<2.8
Sample Analysis 1	<10	84	61.8	87.2	83.6	113	118	95.3
Laboratory Duplicate 1	<10	88.2	67	86.6	87.4	109.6	105	90.4
Relative Percent Difference	4.3%	4.9%	8.1%	0.7%	4.4%	3.1%	11.7%	5.3%
Sample Analysis 2	41				55.2		603	
Laboratory Duplicate 2	42.8				53.6		667	
Relative Percent Difference	4.3%				2.9%		10.1%	
Spike Concentration	200	200	200	200	200	200		
Sample Concentration	37.2	65.8	68	68	42.4	71.8		
Spike Recovery (Percent)	85%	89%	101%	93%	126%	121%		
Time of Incubation (Days)	0	29	57	85	176		358	

Table 9.5   Typical Quality Performance Data for Continuing Calibration Check Standards for Benzene in Water. All Values are µg/L Unless Otherwise Indicated

Date Collected	5/2/01	5/31/01	6/28/01	7/26/01	10/25/01	11/16/01	4/25/02	5/15/03
Date Analyzed	5/31/01	9/24/01	9/3/01	9/21/01	11/30/01	11/26/01	4/25/02	5/16/03
Check Standard Nominal	20	200	200	200	200	200	200	10
Check Standard Measured	22.6	209	204	239	204	228	235	10.1
Percent of Check Standard	113.0%	104.5%	102.0%	119.5%	102.0%	114.0%	117.5%	101.0%
Check Standard Nominal	200	20	200	20	20.0	200	20	50
Check Standard Value	246	21.3	237	21.3	20.4	236	18.1	50.6
Percent of Check Standard	123.0%	106.5%	118.5%	106.5%	102.0%	118.0%	90.5%	101.2%
Check Standard Nominal	200	200	200	200			200	100
Check Standard Measured	216	193	204	233			184	102

Table 9.6 Typical Quality Performance Data for Analysis of Benzene in Water, Including Blanks, Laboratory Duplicates, and Matrix Spikes. All Values are µg/L Unless Otherwise Indicated

Date Collected	5/2/01	5/31/01	6/28/01	7/26/01	10/25/01	11/16/01	4/25/02	5/15/2003
Date Analyzed	5/31/01	9/24/01	9/3/01	9/21/01	11/30/01	11/26/01	4/25/02	5/16/2003
Blank 1	no report	<0.5	<0.5	<0.5	<0.5	<1	<0.3	<0.18
Blank 2	no report	<0.5	<0.5	<0.5			<0.3	<0.18
Sample Analysis 1	604	20.2	13.3	15.9	3.3	<1	10.3	15.6
Laboratory Duplicate 1	562	20.6	15	15.6	2.8	<1	8.45	13.6
Relative Percent Difference	7.2%	2.0%	12.0%	1.9%	16.4%		19.7%	13.7%
Sample Analysis 2	48.2				6.2			
Laboratory Duplicate 2	48.6				5.6			
Relative Percent Difference	0.8%				10.2%			
Spike Concentration	200	200	200	200	200	200		
Sample Concentration	36	25.6	17.7	14.9	7.4	<1		
Spike Recovery (Percent)	121%	99%	113%	119%	114%	116%		
Time of Incubation (Days)	0	29	57	85	176		358	
Holding Time (Days)	29	116	67	57	36		0	
Sterile Water Control 1	24.8	20.4	21.8	21.6	16.0		16.1	
Sterile Water Control 2	25.8	20.2	21.2	21.2	14.8		14.8	
Sterile Water Control 3	24.8	20.2	21.0	20.6	13.1		13.1	
Mean	25.1	20.3	21.3	21.1	14.6		14.7	
Standard Deviation	0.58	0.12	0.42	0.50	1.46		1.50	

Table 9.7   Typical Quality Performance Data for Continuing Calibration Check Standards for Ethanol in Water. All Values are mg/L Unless Otherwise Indicated

Date Collected	5/2/01	5/31/01	6/28/01	7/26/01	10/25/01	11/16/01	4/25/02
Date Analyzed	5/31/01	9/24/01	9/3/01	9/21/01	11/30/01	11/26/01	4/25/02
Check Standard Nominal	85	15	15	15	15	25	10
Check Standard Measured	85.8	14.1	15	15.8	15	28.7	10.3
Percent of Check Standard	100.9%	94.0%	100.0%	105.3%	100.0%	114.8%	103.0%
Check Standard Nominal	15	75	15	75	75.0	25	50
Check Standard Value	15.5	75.6	15.1	78.3	71.7	27.4	49.9
Percent of Check Standard	103.3%	100.8%	100.7%	104.4%	95.6%	109.6%	99.8%
Check Standard Nominal	85				15		
Check Standard Measured	87.2				15.5		
Percent of Check Standard	102.6%				103.3%		

Table 9.8   Typical Quality Performance Data for Analysis of Ethanol in Water, Including Blanks, Laboratory Duplicates, and Matrix Spikes. All Values are mg/L Unless Otherwise Indicated

Date Collected	5/2/01	5/31/01	6/28/01	7/26/01	10/25/01	11/16/01	4/25/02
Date Analyzed	5/31/01	9/24/01	9/3/01	9/21/01	11/30/01	11/26/01	4/25/02
Blank 1	<1	<1	<1	<1	<1	<1	<1
Blank 2	<1	<1	<1	<1	<1	<1	<1
Sample Analysis 1	1980					17.5	37.5
Laboratory Duplicate 1	1950					18.3	37.2
Relative Percent Difference	1.5%					4.5%	0.8%
Spike Concentration	40	40	40	40	40	100	100
Sample Concentration	83.7	83.2	<1	<1	37.4	52	37.4
Spike Recovery (Percent)	104%	102%	104%	101%	94%	93.6%	95.8%

Section 10 -

References -

Amerson, I. and R. L. Johnson. A natural gradient tracer test to evaluate natural attenuation of MTBE under anaerobic conditions. *Ground Water Monitoring & Remediation* 23 (1): 54-91 (2002).

Bradley, P. M., F. H. Chapelle, and J. E. Landmeyer. Methyl t-butyl ether mineralization in surface-water sediment microcosms under denitrifying conditions. *Applied and Environmental Microbiology* 67 (4): 1975-1978 (2001a).

Bradley, P. M., F. H. Chapelle, and J. E. Landmeyer. Effect of redox conditions on MTBE biodegradation in surface water sediments. *Environmental Science & Technology* 35 (23): 4643-4647 (2001b).

Bradley, P. M., J. E. Landmeyer, and F. H. Chapelle. Widespread potential for microbial MTBE degradation in surface-water sediments. *Environmental Science & Technology* 35 (4): 658-662 (2001c).

Borden, R. C., R. A. Daniel, L. E. LeBrun IV, and C. W. Davis. Intrinsic biodegradation of MTBE and BTEX in a gasoline-contaminated aquifer. *Water Resources Research* 33 (5): 1105-1115 (1997).

Buscheck, T. E., and C. M. Alcantar. "Regression techniques and analytical solutions to demonstrate intrinsic bioremediation." In *Intrinsic Bioremediation, Proceedings of the Third International In Situ and On-Site Bioremediation Symposium*, San Diego, CA, (1995).

Cline, P. V., J. J. Delfino, and P. S. C. Rao. Partitioning of aromatic constituents into water from gasoline and other complex solvent mixtures. *Environmental Science & Technology* 25 (5): 914-920 (1991).

Cozzarelli, I. M., and Baehr, A. L. "Volatile fuel hydrocarbons and MTBE in the environment treatise on geochemistry." In *Treatise on Geochemistry: Vol. 9, Environmental Geochemistry,* Amsterdam: Elsevier/Pergamon, 2003, 433-474.

Crumbling, D. M., and B. Lesnik. *Analytical issues with MTBE and related oxygenate compounds. L.U.S.T.LINE: A report on federal and state programs to control leaking underground storage tanks.* New England Interstate Water Pollution Control Commission, Bulletin 36, 2000.

Day, M. J. and Gulliver, T. "Natural attenuation of tert butyl alcohol at a Texas chemical Plant." In *MTBE Remediation Handbook,* Amherst, MA: Amherst Scientific Publishers, 2003, 541-560.

Deeb, R. A., K. M. Scow, and L. Alvarez-Cohen. Aerobic MTBE biodegradation: An examination of past studies, current challenges, and future research directions. *Biodegradation* 11: 171-186 (2000a).

Deeb, R. A., S. Nishino, J. Spain, H. Hu, K. Scow, and L. Alvarez-Cohen. "MTBE and benzene biodegradation by a bacterial isolate via two independent monooxygenase-initiated pathways." Paper presented at the *Division of Environmental Chemistry, American Chemistry Society Symposium,* San Francisco, CA, March 26-30, 2000, (2000b).

Deeb, R. A., H. Y. Hu, J. R. Hanson, K. M. Scow, and L. Alvarez-Cohen. Substrate interactions in BTEX and MTBE mixtures by an MTBE-degrading isolate. *Environmental Science & Technology* 35 (2): 312-317 (2001).

Durrant, G. C., M. Schirmer, M. D. Einarson, R. D. Wilson, and D. M. Mackay. "Assessment of the dissolution of gasoline containing MTBE at LUST Site 60, Vandenberg Air Force Base, California. In *Proceedings of the Petroleum Hydrocarbons and Organic Chemicals in Groundwater: Prevention, Detection, and Restoration. National Ground Water Association,* Houston, TX, November 17-19, 1999.

Finneran, K. T. and D. R. Lovley. Anaerobic degradation of methyl tert-butyl ether (MTBE) and tert-butyl alcohol (TBA). *Environmental Science and Technology* 35 (9): 1785-1790 (2001).

Finneran, K. T. and D. R. Lovley. "Anaerobic *in situ* bioremediation." In *MTBE Remediation Handbook,* Amherst, MA: Amherst Scientific Publishers, 2003, 265-277.

Fiorenza, S. and H. S. Rifai. Review of MTBE biodegradation and bioaugmentation. *Bioremediation Journal* 7 (1): 1-36 (2003).

Fortin, N.Y. and M. A. Deshusses. Treatment of methyl tert-butyl ether vapors in biotrickling filters. 1. Reactor startup, steady state performance, and culture characteristics. *Environmental Science and Technology* 33 (17): 2980-2986 (1999).

Francois, A. L., H. Garnier, F. Mathis, F. Fayolle, and F. Monot. Roles of *tert*-butyl formate, *tert*-butyl alcohol and acetone in the regulation of methyl *tert*-butyl ether degradation by Mycobacterium austroafricanum IFP 2012. *Applied Microbiology Biotechnology* 62: 256-262 (2003).

Garnier, P., R. Auria, M. Magana, and S. Revah, S. "Cometabolic biodegradation of Methyl *t*-butyl Ether by a soil consortium." In *In Situ Bioremediation of Petroleum Hydrocarbon and Other Organic Compounds. The Fifth International In Situ and On-Site Bioremediation Symposium*, San Diego, CA, 1999.

Gray, J. R., G. Lacrampe-Couloume, D. Gandhi, K. M. Scow, R. D. Wilson, D. M. Mackay, and B. S. Lollar. Carbon and hydrogen isotopic fractionation during biodegradation of methyl tert-butyl ether. *Environmental Science & Technology* 36 (9): 1931-1938 (2002).

Hanson, J. R., C. E. Ackerman, and K. M. Scow. Biodegradation of methyl tert-butyl ether by a bacterial pure culture. *Applied and Environmental Microbiology* 65: 4788-4792 (1999).

Hattan, G., B. Wilson, and J. T. Wilson. Performance of conventional remedial technology for treatment of MTBE and benzene at UST sites in Kansas. *Remediation* 14 (1): 85-94 (2003).

Hickman, G. T. and J. T. Novak. Relationship between subsurface biodegradation rates and microbial density. *Environmental Science and Technology* 23 (5): 525-532 (1989).

Hickman, G.T., J. T. Novak, M. S. Morris, and M. Rebhun. Effects of site variations on subsurface biodegradation potential. *Research Journal of the Water Pollution Control Federation* 61 (9): 1564-1575 (1989).

Hristova, K., B. Gebreyesus, D. Mackay, and K. M. Scow. Naturally occurring bacteria similar to the methyl *tert*-butyl ether (MTBE) –degrading strain PM1 are present in MTBE-contaminated groundwater. *Applied and Environmental Microbiology* 69 (5): 2616-2623 (2003).

Hubbard, C. E., J. F. Barker, S. F. O'Hannesin, M. Vandegriendt, and R. W. Gillham. *Transport and fate of dissolved methanol, methyl-tertiary-butyl ether, and monoaromatic hydrocarbons in a shallow sand aquifer.* Health & Environmental Sciences Department, American Petroleum Institute 4601. Washington, DC, 1994

Hunkeler, D., B. J. Butler, R. Aravena, and J. F. Barker. Monitoring biodegradation of methyl tert-butyl ether (MTBE) using compound-specific carbon isotope analysis. *Environmental Science & Technology* 35 (4): 676-681 (2001).

Huskey, W.P. "Origins and interpretations of heavy atom isotope effects." In *Enzyme Mechanism from Isotope Effects*, Boca Raton, FL: CRC Press 1991, 37-72.

Hyman, M., C. Taylor, and K. O'Reilly. "Cometabolic degradation of MTBE by *iso*-alkane-utilizing bacteria from gasoline-impacted soils. Bioremediation and phytoremediation of chlorinated and recalcitrant compounds." In: *The Second International Conference on Remediation of Chlorinated and Recalcitrant Compounds,* Monterey, CA, 2000.

Jensen, H. M. and E. Arvin. "Solubility and degradability of the gasoline additive MTBE, methyl *tert*-butyl ether and gasoline compounds in water." In *Contaminated Soil '90*, Dordrecht, the Netherlands, 1990.

Kane, S. R., H. R. Beller, T. C. Legler, C. J. Koester, H. C. Pinkart, R. U. Halden, and A. M. Happel. Aerobic biodegradation of methyl tert-butyl ether by aquifer bacteria from leaking underground storage tank sites. *Applied and Environmental Microbiology.* 67 (12): 5824-5829 (2001).

Kane, S. R., T.C. Legler, and L. M. Balser. "Aerobic biodegradation of MTBE by aquifer bacteria from LUFT sites." In *Proceedings of the Seventh International In Situ and On-Site Bioremediation Symposium,* Orlando, FL, 2003.

Kolhatkar, R., J. T. Wilson, and L.E. Dunlap. "Biodegradation of MTBE at multiple UST sites." In *Proceedings of the Petroleum and Organic Chemicals in Ground Water: Prevention, Detection and Remediation Conference & Exposition,* Anaheim, CA, 2000.

Kolhatkar, R, J. T. Wilson, and G. Hinshalwood. "Natural biodegradation of MTBE at a site on Long Island, NY." In *Proceedings of the Sixth International In Situ and On-Site Bioremediation Symposium,* San Diego, CA, 2001.

Kolhatkar, R., T. Kuder, P. Philp, J. Allen, and J. T. Wilson. Use of compound-specific stable carbon isotope analyses to demonstrate anaerobic biodegradation of MTBE in groundwater at gasoline release site. *Environmental Science & Technology* 36 (23): 5139-5146 (2002).

Kramer, W. H. and T. L. Douthit. "Water soluble phase oxygenates in gasoline from five New Jersey service stations." In: *Proceedings of the Petroleum Hydrocarbons and Organic Chemicals in Ground Water: Prevention, Detection, and Remediation Conference and Exposition*, Anaheim, CA, 2000.

Kuder, T., R. P. Philp, R. Kolhatkar, J. T. Wilson, and J. Allen. "Application of stable carbon and hydrogen isotopic techniques for monitoring biodegradation of MTBE in the field. In *Proceedings of the 2002 Petroleum Hydrocarbons and Organic Chemicals in Ground Water: Prevention, Assessment, and Remediation, Conference and Exposition*, Atlanta, GA, 2002.

Kuder, T., R. Kolhatkar, J. Wilson, K. O'Reilly, P. Philp, and J. Allen. "Compound-specific carbon and hydrogen isotope analysis-field evidence of MTBE bioremediation." In *Proceedings of the 2003 National Ground Water Association Focus Conference on MTBE: Assessment, Remediation, and Public Policy,* Baltimore, MD, 2003.

Kuder, T., J. T. Wilson, P. Kaiser, R. Kolhatkar, P. Philp, and J. Allen. Enrichment of stable carbon and hydrogen isotopes during anaerobic biodegradation of MTBE–microcosm and field evidence. *Environmental Science & Technology* 39 (1): 213-220 (2005).

Kuder, T., R. Kolhatkar, J. Wilson, P. Philp, and J. Allen. "Compound-specific isotope analysis of MTBE and TBA for bioremediation studies. In *Proceedings of the 2003 Natioinal Ground Water Association Focus Conference on MTBE: Assessment, Remediation, and Public Policy,* Baltimore, MD, 2004.

Landmeyer, J. E., J. F. Pankow, and C. D. Church . "Occurrence of MTBE and tert-butyl alcohol in a gasoline-contaminated aquifer." Paper presented at the *Division of Environmental Chemistry, American Chemical Society*, San Francisco, CA, April 13-17, 1997.

Landmeyer, J. E., F. H. Chapelle, P. M. Bradley, J. F. Pankow, C. D. Church, and P. G. Tratnyek. Fate of MTBE relative to benzene in a gasoline-contaminated aquifer (1993-98), *Ground Water Monitoring and Remediation* 18 (4): 93-102 (1998).

Landmeyer, J. E., F. H. Chapelle, H. H. Herlong, and P. M. Bradley. Methyl tert-butyl ether biodegradation by indigenous aquifer microorganisms under natural and artificial oxic conditions. *Environmental Science and Technology* 35 (6): 1118-1126 (2001).

Lin, Z., J. T. Wilson, and D. D. Fine. Avoiding hydrolysis of fuel ether oxygenates during static headspace analysis. *Environmental Science & Technology* 37 (21): 4994-5000 (2003).

McLoughlin, P. W., R. J. Pirkle, D. Fine, and J. T. Wilson. Production by acid hydrolysis of MTBE during heated headspace analysis and evaluation of a base as a preservative. *Ground Water Monitoring and Remediation* 24 (4): 57-66 (2004).

Mace, R. E., R. S. Fisher, D. M. Welch, and S. P. Parra. *Extent, mass, and duration of hydrocarbon plumes from leaking petroleum storage tank sites in Texas.* Geological Circular 97-1, Bureau of Economic Geology, University of Texas, Austin, TX, 1997.

Mace, R. E. and W. Choi. "The size and behavior of MTBE plumes in Texas." In *Proceedings of the Petroleum Hydrocarbons and Organic Chemicals in Ground Water: Prevention, Detection, and Remediation Conference, API and NGWA Conference and Exposition,* Houston, TX, 1998.

Mariotti, A., J. C. Germon, P. Hubert, P. Kaiser, R. Letolle, A. Tardieux, P. Tardieux. Experimental determination of nitrogen kinetic isotope fractionation: some principles; illustration for the denitrification and nitrification processes. *Plant and Soil* 62 (3): 413-430 (1981).

Mo, K., C. O. Lora, A. E. Wanken, M. Javanmardian, X. Yang, and C. F. Kulpa. Biodegradation of methyl t-butyl ether by pure bacterial cultures. *Applied Microbiology and Biotechnology* 47: 69-72 (1997).

Mormille, M. R., S. Liu, and J. M. Suflita. Anaerobic biodegradation of gasoline oxygenates: Extrapolation of information to multiple sites and redox conditions. *Environmental Science and Technology* 28 (9): 1727-1732 (1994).

National Research Council. *Natural Attenuation for Groundwater Remediation.* Washington, DC: National Academy Press, 2000.

New England Interstate Water Pollution Control Commission. *Summary report on a survey of state experiences with MTBE and other oxygenate contamination at LUST sites.* August 2000, 2003.

Newell, C. J., R. K. McLeod, and J. R. Gonzales. *BIOSCREEN natural attenuation decision support system,* EPA/600/R-96/087, Cincinnati, OH: U.S. Environmental Protection Agency, 1996.

Newell, C. J., H. S. Rifai, J. T. Wilson, J. A. Connor, J. A. Aziz, and M. P. Suarez. *Calculation and use of first-order rate constants for monitored natural attenuation studies*, EPA/540/S-02/500, Cincinnati, OH: U.S. Environmental Protection Agency, 2002.

Novak, J. T., C. D. Goldsmith, R. E. Benoit, and J. H. O'Brien. Biodegradation of methanol and tertiary butyl alcohol in subsurface systems. *Water Science Technology* 17 (9): 71-85 (1985).

Odencrantz, J. E. "Implication of MTBE for intrinsic remediation of underground fuel tank sites." In: *Proceedings of the Petroleum Hydrocarbons and Organic Chemicals in Ground Water: Prevention, Detection, and Remediation Conference. API, NGWA, STEP Conference and Exposition,* Houston, TX, 1998.

O'Sullivan, G., G. Boshoff, A. Downey, and R. M. Kalin. "Carbon isotope effect during the abiotic oxidation of methyl-tert-butyl ether (MTBE). In *Proceedings of the Seventh International In Situ and On-Site Bioremediation Symposium,* Orlando, FL, 2003.

O'Reilly, K. T., M. E. Moir, C. D. Taylor, C. A. Smith, and M. R. Hyman. Hydrolysis of tert-butyl methyl ether (MTBE) in dilute aqueous acid. *Environmental Science & Technology* 35 (19): 3954-3961 (2001).

Park, K. and R. M. Cowan, R.M. "Effects of oxygen and temperature on the biodegradation of MTBE." In *Proceedings of the 213th ACS National Meeting, Division of Environmental Chemistry*, San Francisco, CA, 1997.

Peargin, T. R. "Relative depletion rates of MTBE, benzene, and xylene from smear zone NAPL. In *Proceedings of the 2000 Petroleum Hydrocarbons and Organic Chemicals in Ground Water: Prevention, Detection, and Remediation. Special Focus: Natural Attenuation and Gasoline Oxygenates. API, NGWA, STEP Conference and Exposition,* Anaheim, CA, 2000.

Peargin, T. R. "Relative depletion rates of MTBE, benzene, and xylene from smear zone non-aqueous phase liquid. In *Proceedings of the Sixth International In Situ and On-Site Bioremediation Symposium*, San Diego, CA, 2001.

Pope, D. F., S. D. Acree, H. Levine, S. Mangion, J. van Ee, K. Hurt, and B. Wilson. *Performance monitoring of MNA remedies for VOCs in ground water*, EPA/600/R-04/027. Cincinnati, OH: U.S. Environmental Protection Agency, 2004.

Rittmann, B. E. "Monitored natural attenuation of MTBE. In *MTBE Remediation Handbook*. Amherst. MA: Amherst Scientific Publishers, 2003, 329-348.

Rixey, W. G. and S. Joshi. *Dissolution of MTBE from a residually trapped gasoline source*, API Soil and Groundwater Research Bulletin No. 13, American Petroleum Institute, 2000.

Robb, Joseph and Ellen E. Moyer. "Natural attenuation of benzene and MTBE at four midwestern U.S. sites." In *MTBE Remediation Handbook,* Amherst, MA: Amherst Scientific Publishers, 2003, 561-578.

Salinitro, J. P., L. A. Diaz, M. P. Williams, and H. L. Wisniewski. Isolation of a bacterial culture that degrades methyl t-butyl ether. *Applied and Environmental Microbiology* 60: 2593-2596 (1994).

Salinitro, J. P., C. S. Chou, H. L. Wisniewski, and T. E. Vipond. "Perspectives on MTBE biodegradation and the potential for *in situ* aquifer bioremediation." In *Proceesings of the Southwest Ground Water Conference—Discussing the Issue of MTBE and Percholorate in Ground Water,* Anaheim, CA, 1998.

Salanitro, J. P., P. C. Johnson, G. E. Spinnler, P. M. Maner, H. L. Wisniewski, and C. Bruce. Field-scale demonstration of enhanced MTBE bioremediation through aquifer bioaugmentation and oxygenation. *Environmental Science and Technology* 34 (19): 4152-4162 (2000).

Schirmer, M. and J. F. Barker. A study of long-term MTBE attenuation in the Borden Aquifer, Ontario, Canada. *Ground Water Monitoring and Remediation* 18 (2): 113-122 (1998).

Schirmer, M., B. J. Butler, J. F. Barker, C. D. Church, and K. Schirmer, K. Evaluation of biodegradation and dispersion as natural attenuation processes of MTBE and benzene at the Borden field site. *Physics and Chemistry of the Earth, Part B: Hydrology, Oceans and Atmosphere* 24 (6): 557-560 (1999).

Schmidt, T., M. Schirmer, H. Weiss, and S. B. Haderlein. Microbial degradation of methyl *tert*-butyl ether and *tert*-butyl alcohol in the subsurface. *Journal of Contaminant Hydrology* 70: 173-203 (2004).

Shih, T., Y. Rong, T. Harmon, and M. Suffet. Evaluation of the impact of fuel hydrocarbons and oxygenates on groundwater resources. *Environmental Science and Technology* 38 (1): 42-48 (2004).

Shorr, G. L. and H. S. Rifai. "Characterizing the intrinsic remediation of MTBE at field sites in Texas." In *Proceedings of the Third International Conference on Remediation of Chlorinated and Recalcitrant Compounds,* Monterey, CA, 2002.

Small, M. C. and J. Weaver. "An updated conceptual model for subsurface fate and transport of MTBE and benzene." In *Proceedings of the Petroleum Hydrocarbons and Organic Chemicals in Ground Water: Prevention, Detection, and Remediation Conference. API, NGWA, STEP Conference and Exposition,* Houston, TX, 1999.

Smallwood, B. J., R. P. Philp, T. W. Burgoyne, and J. D. Allen. The use of stable isotopes to differentiate specific source markers for MTBE. *Environmental Forensics* 2 (3): 215-221 (2001).

Smith, C. A., K. T. O'Reilley, and M. R. Hyman. Characterization of the initial reactions during the cometabolic oxidation of methyl *tert*-butyl ether by propane-grown *Mycobacterium vaccae* JOB5. *Applied and Environmental Microbiology* 69 (2): 796-804 (2003).

Somsamak, P., R. M. Cowan, and M. M. Haggblom. Anaerobic biotransformation of fuel oxygenates under sulfate-reducing conditions. *FEMS Microbiology Ecology* 37: 259-264 (2001).

Somsamak, P., H. H. Richnow, and M. M. Haggblom. Carbon isotope fractionation during anaerobic biotransformation of methyl tert-butyl ether and tert-amyl methyl ether. *Environmental Science and Technology* 39 (1): 103-109 (2005).

Squillace, P. J., J. F. Pankow, N. E. Korte, and J. S. Zorgorski. Review of the environmental behavior and fate of methyl *tert*-buty Ether. *Environmental Toxicology and Chemistry* 16 (9): 1836-1844 (1997).

Srinivasan, P., D. F. Pope, and E. Striz. *Optimal Well Locator (OWL): A screening tool to evaluating locations of monitoring wells,* EPA 600/C-04/017. Cincinnati, OH: U.S. Environmental Protection Agency, 2004.

Steffan, R. J., K. McClay, S. Vainberg, C. W. Condee, and D. Zhang. Biodegradation of the gasoline oxygenates methyl *tert*-butyl ether, ethyl *tert*-butyl ether, and *tert*-amyl ether by propane-oxidizing bacteria. *Applied and Environmental Microbiology* 63 (11): 4216-4222 (1997).

Steffan, R. J., S. Vainberg, C. Condee, K. McClay, and P. Hatzinger. "Biotreatment of MTBE with a new bacterial isolate. Bioremediation and phytoremediation of chlorinated and recalcitrant compounds." In *Proceedings of the Second International Conference on Remediation of Chlorinated and Recalcitrant Compounds,* Monterey, CA, 2000.

Suarez, M. P. and H. S. Rifai. Biodegradation rates for fuel hydrocarbons and chlorinated solvents in groundwater. *Bioremediation Journal* 3 (4): 337-362 (1999).

Suflita, J. M. and M. R. Mormile. Anaerobic biodegradation of known and potential gasoline oxygenates in the terrestrial subsurface. *Environmental Science and Technology* 27 (5): 976-978 (1993).

United States Environmental Protection Agency. Use of monitored natural attenuation at superfund, RCRA corrective action, and underground storage tank sites. *Office of Solid Waste and Emergency Response, Directive* 9200.4-17P (1999).

Varadhan, R., Jin-Song Chen, J. T. Wilson, J. A. Johnson, W. Gierke, and L. Murdie. Evaluation of natural attenuation of benzene and dichloroethanes at the KL Landfill. *Bioremediation Journal* 2 (3-4): 239-258 (1998).

White, H., B. Lesnik, and J. Wilson. Analytical methods for fuel oxygenates. *L.U.S.T.LINE: A Report on Federal and State Programs to Control Leaking Underground Storage Tanks. New England Interstate Water Pollution Control Commission Bulletin* 42: 1-8 (2002).

Wiedemeier, T. H., R. N. Miller, J. T. Wilson, and D. H. Kampbell. "Significance of anaerobic processes for the intrinsic bioremediation of fuel hydrocarbons. In *Proceedings of the Petroleum Hydrocarbons and Organic Chemicals in Ground Water: Prevention, Detection, and Remediation Conference: NWWA/API*, Houston, TX, 1995.

Wiedemeier, T. H., M. A. Swanson, J. T. Wilson, D. H. Kampbell, R. N. Miller, and J. E. Hansen. Approximation of biodegradation rates constant for monoaromatic hydrocarbons (BTEX) in ground water. *Ground Water Monitoring & Remediation* 16 (3): 186-194 (1996).

Wilkin, R. T., J. T. Wilson, M. S. McNeil, and C. J. Adair. Field measurement of dissolved oxygen: A comparison of methods.. *Ground Water Monitoring and Remediation* 21 (4): 124-132 (2001).

Wilson, J. T., J. A. Vardy, J. S. Cho, and B. H. Wilson. *Natural attenuation of MTBE in the subsurface under methanogenic conditions*, EPA/600/R-00/006. Cincinnati, OH: U.S. Environmental Protection Agency, 2000.

Wilson, J. T., and R. Kolhatkar. Role of natural attenuation in the life cycle of MTBE plumes. *Journal of Environmental Engineering* 128 (9): 876-882 (2002).

Wilson, J. T. "Fate and transport of MTBE and other gasoline components." In: *MTBE Remediation Handbook,* Amherst, MA: Amherst Scientific Publishers, 2003a, 19-61.

Wilson, J. T. "Aerobic in situ bioremediation." In: *MTBE Remediation Handbook.* Amherst, MA: Amherst Scientific Publishers, 2003b, 243-264.

Wilson, J. T., C. Adair, P. M. Kaiser, and R. Kolhatkar. Role of anaerobic biodegradation in the natural attenuation of MTBE at a gasoline spill site. *Ground Water Monitoring & Remediation* In Press (2005a).

Wilson, J. T, R. Kolhatkar, T. Kuder, P. Philp, and S. J. Daugherty. Anaerobic degradation of MTBE to TBA in ground water at gasoline spill sites in Orange County, California. *Ground Water Monitoring & Remediation* (In review) (2005b).

Wilson, R. D., D. M. Mackay, and K. M. Scow. In situ MTBE biodegradation supported by diffusive oxygen release. *Environmental Science & Technology* 36 (2): 190-199 (2002).

Yeh, C. K. and J. T. Novak. Anaerobic biodegradation of gasoline oxygenates in soils. *Water Environment Research* 66 (5): 744-752 (1994).